Why Aren't We Saving the Planet?

Global warming. Many of us believe that it is somebody else's problem, that it will affect other people and that other people will come up with the solution. This is not true. 'Global' warming is a global problem: it will affect every single one of us and will only be stopped by a huge shift in our individual attitudes and behaviour. Each time one of us switches on a light, reaches for something in a supermarket, gets into a car or bus, or even chooses what clothes to buy, we are making a choice that can affect the environment. We already know that we need to start making better choices for the sake of our natural world, now.

So why aren't we already saving the planet? This book follows one psychologist's mission to find some answers to this question. Challenged by a recently graduated student to use psychology to find the root of the problem, Geoffrey Beattie (an environmental 'unbeliever') begins a personal and life-changing journey of discovery. The reader is invited to accompany him as he uses psychological methods to examine people's attitudes to global warming. Along the way we find the author's own attitudes being challenged, as well as our own.

This ground-breaking book reflects new and innovative research being carried out into how to change attitudes to the environment and how to encourage sustainable behaviour. It is eminently readable and interesting and, as such, should be read by anyone who is concerned about our planet. In fact, you should also read it if you're not concerned about our planet.

Professor Geoffrey Beattie is Head of School of Psychological Sciences at the University of Manchester. His work on sustainability is carried out in the Sustainable Consumption Institute at the university (founded by Tesco). He obtained his PhD in Psychology from the University of Cambridge (Trinity College) and is a Fellow of the British Psychological Society (BPS). He was awarded the Spearman Medal by the BPS for 'published psychological research of outstanding merit'. Geoffrey was President of the Psychology section of the British Association for the Advancement of Science (2005–2006). His paper with Laura Sale on explicit and implicit attitudes to carbon footprint was short-listed for the International Award for Excellence by the *Journal of Environmental, Cultural, Economic and Social Sustainability*.

Geoffrey Beattie is widely regarded as one of the leading international figures on nonverbal communication and has published 16 books, many of which have either won or been short-listed for major national or international prizes. He was the resident psychologist on all ten 'Big Brother' series and has also appeared on a number of television programmes for BBC1, Channel 4 and UKTV Style (including 'Life's Too Short', 'Family SOS', 'Dump Your Mates in Four Days' and 'The Farm of Fussy Eaters').

Professor Beattie's academic publications have appeared in a wide variety of international journals including *Nature*, *Semiotica* and the *Journal of Language and Social Psychology*. He has also written for a diverse range of newspapers and magazines including: the *Guardian*, *The Times*, the *Independent*, the *Sunday Telegraph*, the *Observer*, the *New Statesman*, and *Marie Claire*.

Why Aren't We Saving the Planet?

A psychologist's perspective

Geoffrey Beattie

Routledge
Taylor & Francis Group

LONDON AND NEW YORK

First published 2010
by Routledge
27 Church Lane, Hove, East Sussex BN3 2FA

Simultaneously published in the USA and Canada
by Routledge
270 Madison Avenue, New York, NY 10016

Routledge is an imprint of the Taylor & Francis Group, an Informa business

Copyright © 2010 Psychology Press

Typeset in Century Schoolbook by
RefineCatch Limited, Bungay, Suffolk
Printed and bound in Great Britain by
TJ International Ltd, Padstow, Cornwall
Cover design by Mark Woods

This publication has been produced with paper manufactured to strict
environmental standards and with pulp derived from sustainable forests.

British Library Cataloguing in Publication Data
A catalogue record for this book is available from the British Library

Library of Congress Cataloging-in-Publication Data

Beattie, Geoffrey.
 Why aren't we saving the planet? : a psychologist's perspective / Geoffrey
Beattie. – 1st ed.
  p. cm.
 Includes bibliographical references and index.
 ISBN 978-0-415-56196-9 (hardcover) - ISBN 978-0-415-56197-6 (pbk.)
1. Attitude (Psychology). 2. Environmental protection – Citizen participation.
I. Title.
 BF327.B43 2010
155.9 – dc22
                                    2010001978

ISBN: 978-0-415-56196-9 (hbk)
ISBN: 978-0-415-56197-6 (pbk)

For my children and those that follow on

Contents

Acknowledgements

I would like to thank Tesco for its generous financial support of the new research that forms the basis of this book. The research was carried out under the auspices of the Sustainable Consumption Institute at the University of Manchester, an institute established with the financial support of Tesco. Laura Sale (mainly implicit and explicit attitudes, dissociation and gesture, persuasion and mood) and Laura McGuire (mainly eye tracking and mood and thinking) were two excellent research assistants, who showed commitment and dedication from the start. The section of the book on the memory of my father and the fort he made for me appeared in a slightly different form in *Protestant Boy* (Granta, 2004) and I thank Granta for permission to use it here. The section on the talk at my youth club first appeared in Beattie, G. (2008), 'What we know about how the human brain works', in Lannon, J. (ed.), *How Public Service Advertising Works*, Henley on Thames: World Advertising Research Centre, and I thank the World Advertising Research Centre for permission to use that section here. A short section on flashbulb memories and my inability to remember how my father sounded appeared in *Admap* magazine, 2009, and I thank them for their permission to use this. The four new studies that form the core of this book either have appeared or will appear in the following international peer-reviewed journals, and I thank the journals for permission to use the material here. The journal references are as follows.

- Beattie, G. and Sale, L. (2009) 'Explicit and implicit attitudes to low and high carbon footprint products',

International Journal of Environmental, Cultural, Economic and Social Sustainability, 5: 191–206. (Chapter 6 in this book).

- — (under review) 'How discrepancies between implicit and explicit attitudes on green issues are reflected in gesture–speech mismatches as the unconscious attitude breaks through', *Semiotica*.
- Beattie, G., McGuire, L. and Sale, L. (2010) 'Do we actually look at the carbon footprint of a product in the initial few seconds? An experimental analysis of unconscious eye movements', *International Journal of Environmental, Cultural, Economic and Social Sustainability*. (Chapter 7 in this book).
- Beattie, G., Sale, L. and McGuire, L. (in press) 'An inconvenient truth? Can extracts of a film really affect our psychological mood and our motivation to act against climate change?' *Semiotica*.

All reasonable efforts have been made to contact copyright holders but in some cases this was not possible. Any omissions brought to the attention of Routledge will be remedied in future editions.

Stephen Waldron, Rod Coombs, Richard Seeley and David McNeill all made very useful comments on an earlier draft of this book and I thank them for their input. Lastly, I would like to thank my PA of ten years, Sylvia Lavelle, who has always managed to hold the fort efficiently and effectively (and with an engaging smile), thereby giving me the opportunity to think, write and channel my worry on the topics in the book itself (rather than on all of the other things that I could be worrying about, probably unnecessarily!).

Motivations implicit and explicit

Everybody needs a vista on the world, and this is mine. A bright airy office lit by lamps huddled in every corner of the room, three desk lamps hugging the corners of the desks not covered in paper, six more lamps standing tall and proud with chrome and off-white shades, one with a white paper shade billowing out, giving out a dull glow, lighting the dead misshapen twigs emerging triumphantly out of a chrome bin. They look as if, even in death, they are stretching out for life by the window in this warm cocoon of an office. There is an old-fashioned clock on one wall, which ticks loudly; it looks old, like something from the fifties, chipped and white and blue, the colour of old-fashioned crockery, but it is a modern copy, a cheap copy already with flaking paint, which somehow manages to make the sound of the original. It is loud and regular in its beat, marking out time, but you can only really hear it when you listen carefully, and you notice how tinny it is. It is a modern, bright office, without the normal strip lights, efficacious and efficient; the noise of the computer whirring in the room makes the whole room feel alive. That's the sound I prefer, the sound of activity and life. I like the glow when students enter the room. It's like coming in from the dark and the cold, into the light and the warmth. They always say the same thing, 'It's very cosy in here.' 'I live here,' I say, 'it has to be.'

I look down onto a tree-lined street through the main university campus. I can see a large part of my world, the optimism of the future, students hurrying by, their coats pulled tight against a cold windy Manchester autumn, grey in early October and they still all look happy. I love that optimism of university life, it's all about the future and possibility; anything is possible, any dream, no matter how ridiculous. I can see the offices opposite; it's an administrative block, human shapes, people I don't recognize, working at set tables, administrators in greyish suits, occasionally moving to check some figures at another table, and then moving back to their first positions. The message they send is stability and continuity, that great university machine working endlessly into the future, the hub of learning. In the afternoon I can see the school kids taking a shortcut to Oxford Road, past my famous building where Rutherford split the atom and then later the very first computer in the world was assembled, the building where Baby was born. And I can see the new recycling bins, all nine of them in a neat row with blue or black tops and a bright orange chute thing on the top, just inviting you to recycle and save the planet. They turned up recently; they line the side of the building like sentries, just standing there watching you, on guard.

I sit and look out of my window and watch one man carefully and tentatively approach the bins. He has come prepared, everything has been sorted well in advance and he stands there in the grey drizzle placing each item carefully and neatly into the correct box. The students hurry past, hardly noticing him; I just notice his ill-fitting trousers and his haversack, purple and green, the colours and fashion of twenty years ago, maybe more. His pullover looks tired and recycled, probably from an Oxfam shop. He is a living embodiment of one of my cultural stereotypes – the repressed eco-warrior, on his own little pathetic moral crusade to save the planet. Quite alone in his efforts, my suspicion is that's what he likes most. It makes him feel different and unique. I catch him glancing up at my room: what attracted him to look up I'm not sure, perhaps it was the brightness on this grey afternoon. He looks up not in a challenging way, but in a vacant sort of way; at first, he is just drawn to the light, and then and only then do I see the mild look of disapproval when he sees me sitting there. It's sometimes dangerous to read too much into a fleeting facial expression, but sometimes it's hard not to. That fleeting look said something about the earth's limited natural resources, and finite sources of natural energy; it said something about academics who should know better. It said something about me and him and the gap between us, in a vertical rather than horizontal plane; it said something about moral and intellectual superiority. I hate people looking down at me. I always have. It never provokes change, just a hardening of attitudes, and a host of ready-made rationalisations. I said something under my breath and turned away. For some reason that look of his had made me momentarily angry.

There was a knock on the door. It was one of my recent graduates, Laura Sale, a former student working with me on how the brain sends its complex messages through gestures and speech to other brains during conversation. She came into the room and visibly sighed, looking round slowly and deliberately at each of my lights. 'Do you need all of those on?' she asked. I like her directness. She reminds me a bit of my mother, who would come out with the first thing that was in her head, regardless of the situation. She too had a thing about lights and heat, but for an entirely different reason. Then it was all economic in the grey days of working-class Belfast of the Troubles. 'Turn those bloody lights off,' she would say, 'not a bit of wonder our bills are so big. I'm a widowed woman, we can't afford to put two bars of the electric fire on, or those bloody lights. Get them turned off.' Perhaps that's why I do it, perhaps I'm celebrating the fact that my life has moved on from those days in the streets of Belfast, perhaps I'm signalling my small stake in consumer culture, gratuitously enjoying my days of material possession. I had heard Steve Jones, the geneticist, on the radio, saying that we lived at the very end of evolution and that when anyone asked him what the Garden of Eden was like, he told them to imagine the present. I love the present, I'm embedded in it; I'm part of the action, not like the resentful and superior man with the haversack.

Laura was still looking at me. There was something in that look that I really didn't like. 'What's your problem?' I asked her. 'Do you actually believe in global warming? Have you not noticed that in Manchester winter lasts from

October until May, and that it rains every day? My view is that you should only believe what your senses tell you, and mine are telling me that it's getting colder around here. Perhaps I should have a "Stop global cooling" sticker on my door. That might do the trick. You could join my movement. It might start small but I'm sure that it would grow.'

She didn't take the bait and didn't respond with any anger or any kind of discernible emotion. But that may be temporary, I thought. 'Stop global cooling,' I chanted quietly; 'Stop global cooling,' I said more loudly in that provocative manner I have that really irritates people. She just smiled in that way that women do at men who aren't behaving in a totally mature fashion, at men who should know better. 'But say there was something in the whole thing,' she said. 'You're a psychologist, wouldn't you find it rewarding to try to do something about it?' 'You mean, like the guy with the moth-eaten pullover and the little haversack?' I asked. 'What are you talking about?' she said. 'What pullover? What haversack?' 'Basically, you want me by the bins down there,' I said, 'recycling bits of my very busy life? Checking out who's watching me, showing that I have all the time in the world and the patience and the moral authority to sort all of the crap of my life into little neat piles and then stick them one by one down that bloody orange chute? And you want me to do that slowly enough so that everyone can see what a great guy I am, and not that selfish bastard that some suspect that I really am?'

'No, that's not what I meant,' she said. 'I want you to use your psychology, everything you know to work out what we would have to do in order to make a difference.' I made a ppppfffffff sound at her cheek, a sharp expulsion of air, a primitive rejection of the idea that seemed to do the trick, although the basic onomatopoeia here, which could form the basis for the word 'piffle', probably helped. 'What would the basic principles be?' she asked. I glanced away, breaking any sort of bond. 'Treat it as an intellectual journey if you like. You don't have to believe in it at the start.' She paused. 'But have you ever thought that the reason that you don't believe in it to start with is because the whole thing is so massive that you might not have the psychology to help you? Perhaps you just feel helpless in the face of great challenges? Perhaps this is a classic case of avoidance behaviour by a psychologist who should know better.'

She knew that I would find this confrontational for highly personal reasons. That week I had been on ITV1 providing expert analyses on *Girls Aloud* in a haunted house, then in a TB hospital for *Ghosthunting with Girls Aloud*. 'If you're out there why don't you fuckin' well show yourselves?' Cheryl had screamed into the nothingness with Kimberley perched precariously on her knee. I had said something about the fight or flight response and what happens to the human body and the human brain when it is prevented from fleeing by social or physical constraints, including Yvette Fielding's constant, and well-practised, challenges – 'You're not going to bottle it, are you Cheryl?' – and Kimberley's ample bottom. It was not what I had imagined myself doing with my degrees in psychology.

I went back to staring out of my window. The problem presented now as an intellectual challenge had everything, even I could see that: social identification and the man with the moth-eaten jumper, risk perception and the fact that I love warm, brightly lit offices and that I'm too busy to think too much about the future, attitudes and behaviour and how to change both, the unconscious mind and conscious reflection, the reasons behind behaviour and the way that we can rationalise our actions, beliefs and knowledge, empathy with others in other parts of the world, cynicism and scepticism about commercial involvement, human perception of the world as a small ecosystem or a giant disconnected macro-system, our emotions and our logic, our feeling that we live in the Garden of Eden or at the very end of days, our belief that evolution is over or that a new cultural evolution has just begun, our innermost thoughts that we can do something or that in the end we can do nothing. Laura smiled back at me. It was the face of optimism. 'Okay,' I said, 'let's think about what psychology might have to offer.' It was a conversational opener, to keep her on board, nothing more. 'Where would I start?' She came back that afternoon with the first paper for me to read and placed it neatly on the side of my desk next to one of my lamps so that it wouldn't be displaced. She also placed it with the words facing me so as to minimise my effort, to cut down my excuses as to why I hadn't time to read it. That was her little bundle of unconscious messages.

The paper was predictable enough in its content and tone. It was the doomsday scenario paper: I am sure that you can imagine the tone. I read it carefully, but embarrassingly it did nothing for me or rather it didn't do what the author clearly thought that it was going to do. She came back later and sat over in the corner of my room while I finished reading it, occasionally looking up, as if I could not be trusted to finish the job in hand. The arguments in the paper made some logical sense, as far as I could see, but the problem was that I was no expert in the field and I felt overwhelmed by the insistent, relentless arguments. It might have all been true, every single word of it, but then again none of it might have been true. It was hard to tell.

But the real problem as far as I could see wasn't the logic or lack of logic or even my ability to discern logic in action, it was in my emotional response to what I was reading. I felt no fear or nowhere near the level of fear that the smug git of an author guessed that I would be feeling. And there was something else. It was as if the article didn't concern me and my behaviour: it was almost as if the article wasn't about me or, dare I say it, my planet. It was written for other people, living less busy lives, with time to reflect and find the recycling bins, and time to grade their rubbish into neat piles, with time to walk to busy appointments, instead of running from their cars, and time to browse in supermarkets and make green considered choices instead of running up and down the aisles at five to ten with the assistant with the bad skin shouting that the shop was already closed, and me shouting back that there was at least another three and a half minutes before closing time and what was his problem.

I needed some emotional response to galvanise me into action. Ask me about what time supermarkets close and who makes that decision and I will give you an emotional response, ask me about the convenience of car parking by my department, and why we can't park just outside, and you will be able to read my visceral response from thirty feet, but ask me about the environment in 2050 or test my galvanic skin response to that iconic image of the polar bear stranded on the raft of ice as it floats away from the polar ice cap and I will give you nothing. Perhaps I don't have the imagination or perhaps I'm too good at thinking up alternative scenarios. Perhaps I have learned to look on the bright side of life. After all I did run a 'happiness' course on 'Richard and Judy' teaching random members of the public who were a bit miserable to be a bit happier. Perhaps, like these slightly miserable people (after the course, that is), I have learned to prime my positive memories; perhaps like them I have learned to see and remember the best bits in any situation. Perhaps, like them, when confronted with the image of the polar bear on the raft of ice I have taught myself to remember immediately that polar bears are dangerous, unpredictable predators, and perhaps like them I think that a stranded polar bear is a safe polar bear, and that its pristine white coat is filthy close up, its fur matted with blackened seal blood and the grey debris of melted ice and gravel, and that the bear only looks cute from a safe distance, preferably a quarter of a mile or more and that little in this image is how it first appears.

'Well, what do you think?' she asked eventually and with more than a hint of expectation. I cleared my throat gently, making time, ready to be vague in my reference, prepared to feign my enthusiasm. I wanted to feel emotional, I wanted to feel fear but I couldn't. But I did feel something, and that was a curious empty feeling inside, accompanied by this genuine intellectual curiosity about how many other people out there were just like me, sitting at their desks murmuring about the need to save the planet and exclaiming about what a terrible mess we had got ourselves into and really deep down inside feeling virtually nothing. That almost produced just a flicker of anxiety; an anxiety about the fact that I clearly wasn't getting the message combined with this odd thought as to whether there might actually be something in it. Just that single thought, 'what if . . .?'. But, I suspected, many people were not really getting the message. Every politician, and journalist and good citizen was lining up on the deck to display their green credentials and publicly announce their fears and anxieties and I was just sitting there at the back of the poop deck doing nothing while the ship went down, or jolted and rocked on its normal crossing. The public proclamations were just too on-message for the likes of me: I wanted to know how everyone really felt.

But then I had a strange and unexpected moment. A sort of momentary intense fear of not feeling fear; a fear of something that was absent, like noticing that my clock had stopped ticking; a brief fear of my emotional stillness coupled, I have to say, with this odd desire to know why I was the way I was, and whether I was alone. Was I really this uncaring human being who didn't give a toss about his children or his environment, including his house overlooking the beautiful moors outside Sheffield – the moors there for hundreds of thousands of years, now with golden brown heather – or his legacy? I had talked to the ex-Formula 1 driver Eddie Irvine the day before for a BBC documentary about Blair Mayne, the co-founder of the SAS with David Stirling, and the living embodiment of the regiment in the Second World War. Eddie Irvine and I had discussed Mayne's coolness under fire, his emotional detachment, his apparent lack of guilt after the war about his combat missions in which he had become the most decorated soldier in the British army for his close-quarters killing. Eddie Irvine like Blair Mayne hailed from Newtownards, a stone's throw from Belfast, and he opened up about his own emotional detachment from aspects of life and the way that images of the Troubles in that small part of Ireland never really troubled him, 'unless there were children involved. Mutilated adults just don't have any real effect on me. And when it comes to Formula 1 I never really cared if I had to manoeuvre another driver into the wall. It just doesn't affect me. In my view it's all about evolution and the survival of the fittest.' He thought nothing of Mayne's lack of guilt or remorse or his emotional detachment from the events unfolding around him. So, what if I had that same kind of emotional detachment and it was that something which was missing in me which was leading me to be so uncaring and unthinking about the environment? What if I should have been doing something, but wasn't?

I love our planet; I love the quiet, rugged moors near my home outside Sheffield, and my runs of nine or ten miles through them with dusk approaching on cold midwinter afternoons, alone, mud-splattered, icy fingers, but with my heart pounding, looking out on a primitive landscape that seems to take on my noisy pulse, as if I had projected it somewhere else. The moors feel alive and fiercely robust, in fine health, fit and well for millennia, and there is me in the middle of it all, just the visitor, just temporary, just passing through. I am allowed temporarily to view its great rugged beauty. I never doubt where the power lies in all of this: perhaps that's why I don't run around trying to save the planet. I love mountain ranges far away; I love snow without footprints, unsullied and pure, and trails that wind to infinity that hint of adventure and maybe spiritual enlightenment at the base of the clouds. I love the feeling of history and permanence. But I have an image in my head of nature more potent than some wild polar bear with filthy legs and blackened paws, and that is an image from Nanda Devi in the Himalayas, a picture taken way above the snow line. White peaks touching the sky above the goddess of joy. And there in the foreground a suntanned face with white goggle marks around the eyes and a fresh growth of beard, a westerner smiling for the camera, like a tourist above the world, pale skin touched by the sun, reddened in parts, an unmistakable joy in his face, touched by the goddess of joy, that unmistakable smile, the eyes crinkled into life, the sign of a human being testing himself in the wilderness, in tune with life, in tune with nature, in tune with himself. This image is my brother and I have another image of him as well, an image of him wrapped in something green and filthy, the kind of thing that climbers would have to hand, that they might dump at the end of the day's climbing, or at the end of an ill-planned expedition, and I can see the rough contours of his body through the grimy, green tarpaulin, his face and body covered, just lumps and bumps visible on the stretched plastic, as he was laid to rest a day after the photo-graph, under some filthy stones with his name scratched on a rock like a warning to fellow travellers in this remote and dangerous region, like Coleridge's Mongol king.

> I would build that dome in air
> That sunny dome! those caves of ice!
> And all who heard should see them there
> And all should cry, Beware! Beware!
> His flashing eyes, his floating hair!
> Weave a circle round him thrice
> And close your eyes with holy dread
> For he on honey-dew hath fed
> And drunk the milk of Paradise.

Beware, beware, that's what that pile of stones said, about someone who never had any illusions, those bright, flashing eyes closed for ever, and that was my emotional legacy. A legacy of loss, a legacy of recognising that everything human is temporary and transient, and it is only kings and politicians, for quite different reasons, who seem to think otherwise; and a realisation that everything you hold dear will pass, no matter how much you want it to stay for ever. Perhaps there is the danger that I am emotionally blunted by life, and the small unpredictable events that have shaped me, and perhaps I am emotionally disempowered by this whole cumulative experience. And perhaps I have a feeling, deep down and buried inside, that nature is a dangerous and unpredictable force and much stronger than we can imagine, and that, when it comes down to it, it can bloody well look after itself without the help of you and me.

I have another radiant image of my brother in mountains, but it is a curious image because he is not physically present in the scene; rather it is an image of an object glistening by a river, like a totem, something that represented him. Once he invited my girlfriend and me to Chamonix for a summer of climbing in the Alps but when we got there eventually, after many mishaps, he had gone. He was too easily bored to wait for us. He had gone somewhere else to climb; his fellow climbers said that they knew he had gone but they just didn't know where he had gone to. We had no tent and no money and I spent the first day wandering aimlessly around the town and the surrounding fields for somewhere dry to sleep with a seventeen-year-old who was terrified of any spider that scuttled, any daddy-long-legs that flapped, or anything with or without legs that could crawl up her body at night. It was never going to be easy to find sanctuary there. But I tried and eventually I found a beautiful bubbling Alpine stream that jostled its way under an ancient moss-covered bridge. The view of the still snow-covered Alps in the background was spectacular, the water was pure, a bridge afforded shelter from the wind and any rain. I was ecstatic. My girlfriend had big doleful grey eyes that looked out from under a black fringe. She sat on the dry grass in her flimsy summer dress and started to sob gently into her hands. She said that she never wanted to go anywhere with me, ever again. She made me promise to start sweeping the river bank for spiders (like the bomb squad from our native Belfast, I had to secure the area inch by inch). Every time I found a spider, like a magician, I made it disappear. She was sobbing so much, with the tears blinding her eyes, she was easily fooled by my sleight of hand, as the small black spider balls were flipped to the floor. I worked my way from where we sat to right under the bridge and there it was on its own as if it had been arranged deliberately and carefully, like an iconic work of art in an exhibition, full of symbolic significance. It was a tin of Heinz baked beans, the metal at the top still shiny, with a ragged top, hastily and hungrily opened. This was the sign that I had been looking for. Others had slept there, English climbers (you just knew that they would be climbers): it would be a safe place to stay.

Some might try to criticise climbers, who are allowed to get so close to the mesmerizing beauty of nature, for soiling the environment in this way. But I had no such feeling of revulsion. I needed a sign and that was it. We slept there that night and the grey-eyed girl felt secure because others had been there before. The next morning there was no sign of the tin can: the stream had probably pulled it into its journey. Months later when my brother and I finally met I discovered, by accident, that it was he who had enjoyed the beans on his first night in Chamonix and rather than pitch a tent he had decided to rough it for one night. He loved the mountains but, like many climbers, he left debris behind: he knew that the mountains could take care of themselves, that nature always triumphs. The debris of human life is always swallowed up.

But what if I have too egocentric a view on our world? What if I am too analytic about my own limited experiences? What if my inaction was my own fault, all down to one or two moments that I had experienced in my life etched on my unconscious mind? It was maybe that feeling plus a certain intellectual curiosity, in which I clearly needed to reassure myself that I was conventionally normal, that galvanised me to do something, to test who believed what and whether or not this would ever line up with their actions. Laura might have been the emotional believer, maybe a catalyst (maybe not); I just wanted to understand why people like me, and there must be many, were doing nothing. It was as simple and as complex as that. But I knew that this was going to be a journey. Like any psychologist I spend the vastly greater part of my time in very familiar terrain, but for this journey I was going to have to travel through some very unfamiliar territory. If psychology was going to offer anything here, by way of explanation, I was going to have to rethink many old assumptions, to retrace steps that I had already taken to get to where I now stood, to look again at many old issues afresh, to climb many new mountains, some unpredictable and treacherous.

'So,' I said, with a good deal more enthusiasm, to Laura still standing there, waiting for my response, 'I'll start at the most basic level, with the individual and his or her basic thinking. Hey, I'm a psychologist, where else would I start?' And we both laughed in the way that people do when they think that they're communicating openly, but know in reality that they have a great deal that they're not yet ready to share.

PART I

Notes on attitude

2

Small things can make a difference

In 2008, in a book entitled *The Hot Topic*, Gabrielle Walker and David King expressed the following important sentiment:

> It's easy to believe that global warming is somebody else's problem – other people will suffer and other people will come up with the solution. However, this is far from the truth. There's a clue in the name: 'global warming' is a truly global problem. None of us is safe from its effects (although some of us have a better chance of adapting to them). We are all part of the problem, and each of us will need to be part of the solution . . . Thinking this way presents the human race as one massive blob. But in fact it's as individuals that we live our lives and make our choices. Every time each of us switches on a light, reaches for something in a supermarket, gets into a car or bus, chooses what clothes to buy or which movie to see, we have all made a difference to the way the economy works. Choices like these have driven the world's economies ever upwards in the twentieth century. They have also led to spiralling greenhouse gas emissions. Now we will all have to adapt our choices to the new realities of the twenty-first century. (2008: 238)

As a psychologist, I find this argument not just persuasive but attractive. It empowers me and my profession. Of course, I agree that it is 'as individuals that we live our lives and make our choices', but I also believe that there is a good deal of complex psychology underpinning the actual behaviour of making choices, choosing one product rather than another, or, indeed, choosing whether to buy a product at all, and maybe putting it back on the shelf (sometimes the hardest choice of all for many people). Human beings have all kinds of predispositions to select one thing rather than another based on their underlying attitudes and beliefs, their habits, what they think others might do in the same situation and a host of other factors, some personal, some social; and some specific, others quite general. I wanted to rethink some of these influences and take some less traditional approaches to this whole problem.

One of the less traditional features of this book is the inclusion of me and my own particular predispositions and peculiarities in the analytic equation. Why would I bother to do this? Is it just hopeless vanity or maybe, just maybe, something else? I hope that it is the something else that is driving this, the fact that I see myself as part of the problem, no better nor worse than many others, people out there like me, who should be doing more for the planet, but aren't. If any of the psychological constructs that I conjure up cannot illuminate me and my particular inertia then they may well fail with others. The inclusion of my own predispositions as an object of study could be both illuminating and perhaps also a little disconcerting, at least to me. Not so much 'physician heal thyself', but psychologist take a long hard look at yourself for once, step back and think.

So I start with myself and my own consumer choices, with shopping at the most basic and mundane level, the kind of shopping that I really don't like (unlike clothes shopping, which I love). Our everyday consumption was killing the planet, we were told, so I wanted to start among that mundane everyday process of people making apparently mindless decisions with hugely significant consequences, according to all the soothsayers. Some of my own research in psychology has focused on the individual and how they read the verbal and nonverbal messages that surround them. So it was no real surprise that I wanted to start my quest here with the decisions individuals make when they wander around supermarkets, making decisions about what to buy and what not to buy on the basis of the marketing messages laid in front of them, without much conscious reflection.

Some products were now appearing with the carbon footprint clearly labelled; the green issue was everywhere, even on the orange juice. So it gave me a chance to see what people thought, what they cared about, how they felt emotionally, and whether deep down inside people really cared about their environment. And I carried out my first experiment on myself, of course: each day for a fortnight I studied the contents of my own carrier bags when I got home from my local Tesco Express for evidence of my own environmental sensitivity. It was a strange sort of experiment, hardly double blind, more than a little biased, for I knew what I wanted to find. Or maybe, in reality, it was a little more open than that. I could after all have surprised myself with my care and sensitivity. But I didn't. I would tip the goods onto my bed and search for the carbon footprint icons, on the back or the sides or wherever they were hiding on the orange juice or the detergent, hoping that somehow unconsciously and unwittingly I had noted these before my hands, without any measurable reflection, snatched the goods from the shelves. What I found was disappointing in the extreme (the fact that I had so many plastic carrier bags to go through in the first place should have held a clue). But was I alone? And if I wasn't buying the low-carbon-footprint orange juice, why wasn't I? And who was? Was any of the whole carbon labelling approach making a difference?

The philosophy behind carbon labelling always seemed logically to me to be a good one. It puts the emphasis on the ordinary consumer and essentially empowers them. The logic runs something like this – climate change is down to consumer excess, so the solution must lie in changes in consumer behaviour. The underlying assumption is that the ordinary consumer is motivated to change and that all they need is the right information to make an informed decision. It all sounds simple enough but there are many assumptions in the reasoning, which any psychologist would be keen to point out. And there were other problems as well. Paul Upham at the Tyndall Centre at the University of Manchester had been conducting research which had demonstrated that in many ordinary people's eyes, carbon labelling was still quite controversial (at least in the period 2008–2009, when he was conducting the research). Severe doubts were being expressed in the interviews that he was conducting about how to measure it, and there were even calls from some organisations to delay the introduction of carbon labelling until there was greater consensus on issues of measurement. There was also considerable cynicism about why certain companies were engaging in it, with people assuming that carbon labelling was more to do with selling products than reducing carbon emissions, and there were doubts as to how effective carbon labelling might actually be. One statistic that jumped off one of his original reports was that it would take thirty-two years of a family drinking low-carbon-footprint orange juice to equal the same family of four flying to Malaga once for their holidays.

So, given the current state of knowledge and desire about carbon labelling, why bother with it at all at the present time? One of the most basic answers would be that it increases the salience of environmental sustainability as a purchasing decision and therefore it empowers consumers to do something – even a little. Ramchandani writing in the *Guardian* in 2006 made the 'Every little helps' point (especially ironic given that Tesco was now funding Paul Upham's research through the Sustainable Consumption Institute at the University of Manchester). Ramchandani wrote: 'A slogan that was written to articulate value, quality and convenience in a multitude of sectors now shows an astonishing fluency in environmental responsibility ... If a few of us do a little – recycling a few carrier bags, say – then every little helps a lot.'

Now the normal way that this slogan is interpreted in the context of sustainability is exactly the way that Ramchandani interpreted it. If we all do a little bit then the sum total of our efforts will mean that a lot is done. But there is another way in which this slogan can be interpreted from a more psychological perspective. Depending on your underlying beliefs, even doing a little might be very important because if you don't act in accordance with your underlying beliefs this may set up an internal state of discomfort called 'cognitive dissonance' (Festinger 1957) within you. People like to avoid being in this state and they will, on occasion, change their underlying beliefs to match the behaviour that they are engaged in. (According to Festinger we have some kind of inner drive to hold all of our attitudes, beliefs and behaviours in harmony and essentially to avoid disharmony or dissonance.) So this means that if the majority of individuals are, in terms of their underlying attitude, very pro-environment then we should allow them to do something, like buying low-carbon-footprint orange or not taking the family to Malaga for a long weekend, or they might well change their underlying attitude. What this line of argument highlights is that underlying belief is critical to the whole process of carbon labelling because it highlights the dangers of not allowing people in their everyday life to do something that they do actually believe in.

I started to put myself into the equation here, as I vowed to. There is always the possibility that I was very pro-green in terms of my underlying values at some point in the not so distant past, but maybe because of my hectic lifestyle ('everything at the last minute, that's you,' my mother would say), I was always having to make last-minute consumer choices (poorly considered, hasty, driven by dubious judgements about 'value' rather than anything else), many of which were simply not in accordance with my underlying values. So maybe I had been in a state of cognitive dissonance for some time without really noticing it, and perhaps this uncomfortable nagging little state drove me to change my underlying values so that they now matched my behaviour (less green now, more selfish). It was at least a scenario (but maybe not that plausible: at least, it didn't feel plausible).

Of course, underlying beliefs are also important in a more fundamental way because they determine what carbon labelling is all about. Is promoting greener consumer choice just about giving information to customers whose underlying attitude is already pro-low carbon or do we actually have to work on changing the attitude as well in order to promote green consumer behaviour?

My psychological journey, therefore, had to start with at least some of these abstract concepts that lie behind behaviour (words like 'belief', 'underlying values', 'core attitudes' – terms that trip off the tongue so easily but may be a little more complex in reality than we ordinarily assume). The obvious one to begin with was the concept of 'attitude', one of the central concepts of psychology, indeed one of the pillars of psychology, almost since the very beginning. Writing in 1935 in one of the classic books on social psychology, Gordon Allport stated that 'The concept of attitude is probably the most distinctive and indispensable concept in contemporary American social psychology. In fact several writers . . . *define* social psychology as the scientific study of attitudes' (emphasis in original). (Allport 1935: 784)

Perhaps the single most significant contributor in the evolution of the attitude construct was Allport himself. As one of the founders of social psychology, Allport made a number of significant contributions, including the development of a new sort of theory of personality, a trait theory, which rejected both the psychoanalytic and behavioural approaches to personality description and analysis. Allport was repelled by the way that psychoanalysis dug far too deeply for the root causes of human action, while the latter approach, he thought, just (literally) skimmed the surface and, in his view, did not go nearly deep enough. It was Allport who gave early social psychology much of its distinctive feel, and led it carefully away from where it might have ended up in the endless psychoanalytic depths.

What is fascinating about Allport, the man who has steered social psychology on its course for the past seventy-five years (even after his demise) is that he liked to give us glimpses into his own life (so maybe some aspects of my 'non-traditional' approach in this book are not that novel), so that we might understand why he chose one scientific course rather than another. And there is one particular autobiographical nugget that stands out from all the others. As a student, he tells us, he visited Freud in Vienna, and that one chance visit changed him and changed the future course of psychology. What happened there was basically that dramatic (in a non-dramatic, highly personal sort of way). He liked to tell the story of the encounter as evidence of the psychoanalytic tendency to read too much into everything, without first considering the more obvious and more parsimonious explanation. And this story has been repeated and reproduced many times. Over the years, it has moved away from being a private story of an embarrassing moment in a young man's life to become something of a parable about psychoanalysis and the emergence of the science of social psychology and its retreat from what Allport himself liked to call 'psychoanalytic excess'. But the events described were clearly critical in Allport's development and therefore were critical to the development of the discipline of social psychology.

It begins with Allport's visit to Austria in 1920. Allport had arranged a meeting with Freud in his office in Vienna. Freud was at the height of his fame, and Allport was then just twenty-two years old, having recently graduated with a Bachelor's degree from Harvard but nevertheless with the forwardness and confidence, no matter how shaky, to write to the great man to set up the visit. In Allport's own words: 'I wrote to Freud announcing that I was in Vienna and implied that no doubt he would be glad to make my acquaintance. I received a kind reply in his own handwriting inviting me to come to his office at a certain time.' On entering Freud's office, Allport was greeted with the familiarity of the room, known even then, and an expectant but uncomfortable silence that seemed to open up and engulf them both. Here he was in front of the great man, but he found himself just staring down at the red-patterned Berber rug in the famous inner office, the matted walls painted deep red, the walls laden with pictures of dreams with all their iconic and provocative symbolism, and fragments from antiquity buried deep in the earth for centuries now released like suppressed memories brought back into consciousness, the whole room reeking of decayed cigar smoke. The heady, stale smell of success that almost made Allport choke.

Allport coughed briefly. The bookshelf behind Freud's desk acted as a reminder to all who entered the room of Freud's own great intellectual journey, with books by

Goethe, Shakespeare, Heine, Multatuli and Anatole France, the dramatists, philosophers and poets, who had recognised the power of the unconscious. Allport noted each of these books in turn: he had never heard of a number of the authors; they were outside his realm. He felt intimidated. And in that silence when he dared to lift his head, he had the opportunity to scrutinise some of the other pictures hanging around the room, including *Oedipus and the Riddle of the Sphinx*, where Oedipus stands, wearing a traveller's cloak with a petasos cap hanging over his shoulder, addressing the woman-headed winged lion. Oedipus' right hand is extended in gesture, open and dynamic. Allport could see the famous couch covered with velvet cushions and a patterned Qashqa'i rug with the three linked octagons symbolising, to some of Freud's followers at least, the uterus contractions during parturition. But the three linked octagons merely acted as a reminder to Allport of the id, the ego and the superego and the holy trinity of the psyche, and the neat packaging of psychoanalytic ideas into a list of three. A list of three like some cheap advertising slogan that sells all products, 'A Mars a day helps you work, rest and play', or 'Coke – delicious, wholesome, thirst quenching', first coined in 1909, but certainly around in Allport's time. Allport had noticed that there are always three in the list, always three, when you want to sell big time that is (and 'wholesome' indeed!). And Allport noticed the plush green armchair where Freud would sit behind his patients while they engaged in free association. Allport was drawn to the expression of Doctor Charcot in the famous painting by André Brouillet of Doctor Charcot at work at the Salpetrière, with the hysterical female patient in full seizure displayed before the staff and medical students. Freud himself had been a student in the audience many years before and the painting may have reminded him of those happier carefree days or it may have acted as a symbolic reminder of the power of the mesmeric great teacher who had pioneered the use of hypnosis, and the effects he was having on his enchanted and captivated audience and on the frozen, hysterical female patient who was helpless in front of them all, with only the great Doctor Charcot or Freud himself capable of understanding her malady. It made Allport uncomfortable in the extreme; it was all a bit too *showy* for him; it went against his own implicit beliefs.

In the days before Harvard, Allport had been a shy, studious boy, often teased by his school friends for having just eight toes as a result of a birth defect. He had a veneer now of Harvard sophistication, of the new international academic in the making, but silences like this made him more uncomfortable. It reminded him of who he had been; maybe of who he was. He knew better than most that personality never really changes. He needed to say something, so he thought that he would make an observation, a psychological observation of something that he had just witnessed. He described how he had watched a small boy of about four on the tram car on the way to Freud's office who was terrified of coming into contact with any dirt. The boy refused to allow a particular man on the tram to sit beside him because he thought that the man was dirty, despite his mother's cajoling and reassurance. Allport studied the woman in question and noted how neat and tidy she was, and how domineering in her approach to her son. Allport hypothesised that the dirt phobia of the young boy had been picked up from his mother, someone who needed everything neat and tidy and in its correct place. 'To him [the boy] everything was *schmutzig*. His mother was a well-starched *Hausfrau*, so dominant and purposive looking that I thought the cause and effect apparent.'

Freud looked at Allport carefully for the first time, with his 'kindly therapeutic eyes', and then asked, 'And was that little boy you?' Allport blinked uncomfortably and said nothing, appalled by Freud's attempt to psychoanalyse him on the spot. Allport himself *knew* that his observation was driven by the desire to fill the silence, his desire to display to Freud that as a psychologist he, the young man from Harvard, never stopped observing, and his desire to connect with the great man, maybe even his need for belonging through this essential connection. These were all manifest and clear motives, maybe at different levels but all open to the conscious mind, which should be obvious to all. What it was not was any unconscious desire to reveal his own deep-seated uncertainties and anxieties resulting from problems in potty training back in Montezuma, Indiana. Allport tried to change the subject but the damage had been done. 'I realized that he was accustomed to neurotic defenses and that my manifest motivation (a sort of rude curiosity and youthful ambition) escaped him. For therapeutic progress he would have to cut through my defenses, but it so happened that therapeutic progress was not here an issue' (Allport 1967: 7–8).

Allport later wrote that the 'experience taught me that depth psychology, for all its merits, may plunge too deep, and that psychologists would do well to give full recognition to manifest motives before probing the unconscious'. This was a clear example, in his mind, of the 'psychoanalytic excess' that he liked to detail, although needless to say psychoanalysts ever since have not necessarily been convinced by his argument. Faber, writing in 1970, suggested that Freud got it exactly right, that Allport had 'practiced a kind of deception in order to work his way into Freud's office. The deception lay in his implied claim that (1) he genuinely wanted to meet Freud as a human being and as an intellectual rather than as an object, and (2) that he (Allport) himself was worth meeting as a human being and as an "intellectual" ' (Faber 1970: 62). Faber believes that Freud saw through this deception quickly and that by asking Allport whether he was the little boy in the story he was in fact indicating to Allport that he knew that he was a 'dirty little boy' and that by putting this question to him, Freud was merely trying to restart the conversation in an honest and straightforward way. Elms (1993) attributed even greater analytic power and clarity of thinking to Freud in this meeting. Allport's childhood was characterised by 'plain Protestant piety' (Allport 1968), with an emphasis on clear religious answers to difficult theological and personal questions and an upbringing in an environment that doubled as a home and as a hospital and that was run by Allport's physician father. According to Elms, the question had such a marked effect on Allport because he 'was still carrying within him the super-clean little boy' brought up in that literal and metaphorical sterile Protestant environment where patients were to be avoided as sources of infection and possible danger.

But Allport was convinced of his own explanation for the event and was determined to do something about this psychoanalytic excess. This meeting encouraged Allport to develop something different, a different sort of approach to the human mind, an approach that stayed with us for some sixty years before anyone really dared challenge it in a systematic way. An approach based around conscious reflection and the power of language to uncover and articulate underlying attitudes, to bring attitudes into the open where they could be studied and analysed objectively and scientifically. This was to characterise the new social psychology that held sway for the next half-century or more and gave us our core methods and techniques in social psychology. This is the armoury that most psychologists who are interested in doing something about climate change necessarily draw on.

3

Measuring attitudes to sustainability: easily, consciously and wrongly?

Allport made genuine advances in many areas of psychology. In order to develop his science of personality, He began by going through the dictionary and identifying every single lexical item that could be used to describe a person. His trawl pulled in 4500 trait-like words. In these lexical descriptors, the words used in everyday life, he saw the start of a new scientific theory of personality, rooted in the stuff of everyday life, in the words that we use consciously and deliberately to describe other people. It was four years later, in 1924 at Harvard, that Allport began what was in all likelihood the very first course on personality in the United States – 'Personality: Its Psychological and Social Aspect'. It was the kind of course that could well fit into the modern psychology curriculum. He went on to develop theories and write books on prejudice, the psychology of rumour and the concept of the self, developing the careers of many outstanding social psychologists including Jerome Bruner, Stanley Milgram, Leo Postman, Thomas Pettigrew and M. Brewster Smith. Another of his students was Anthony Greenwald, whom we shall be encountering in a subsequent chapter. Given Allport's stance on Freud's fixed attentional gaze on the unconscious, it is highly ironic that Greenwald is best known for taking one of Allport's core concepts, the attitude, and detailing the unconscious or implicit aspects of it; indeed, he challenged the whole basis for identifying and measuring it, but more of this later.

Allport himself viewed the concept of the attitude as the central plank of the new psychology. He defined it as 'a mental and neural state of readiness organised through experience, exerting a directive or dynamic influence upon the individual's response to all objects and situations with which it is related' (1935: 810). In other words, an attitude is an internal state of mind affected by what we do which affects our behaviour towards the world around us. In 1935 Allport announced proudly that this concept of the attitude was social psychology's 'most distinctive and indispensable concept' (1935: 798). Its importance should be clear – it should have a major impact on our behaviour (but of course it's not the only factor and, in 1991, Azjen argued that the subjective norm, or how you think others will behave, and the level of perceived behavioural control, in other words the control you have over the particular behaviour, are also crucial).

Of course, there is nothing in Allport's formal definition that formally excludes a possible unconscious component to the attitude (see also Greenwald and Banaji 1995: 7). Indeed, Doob, another great scholar of attitude, working in the years shortly after Allport, had defined an attitude as 'An implicit, drive-producing response considered socially significant in the individual's society' (1947: 136) and in 1992 in an article in which he looked back at the early development of social psychology, he wrote that in the 1940s and earlier the notion of attitudes operating unconsciously was quite acceptable to many researchers.

But when you read this early work of Allport with fresh eyes, I think that you could go much further than this. I think that in Allport's classic (1935) chapter, and despite what happened to him in Freud's office, he initially displays considerable awareness of the unconscious dimensions of an attitude and great sensitivity to this unconscious aspect. When he talks about the early German experimentalists from the Würzburg school, he points out that they believed that attitudes were

> neither sensation, nor imagery, nor affection, nor any combination of these states. Time and again they were studied by the method of introspection, always with meagre results. Often an attitude seemed to have no

representation in consciousness other than a vague sense of need, or some indefinite or unanalyzable feeling of doubt, assent, conviction, effort, or familiarity. (Allport 1935: 800)

Some psychologists (like Clarke 1911) clearly thought that attitudes were represented in consciousness through 'imagery, sensation and affection', but Allport himself seemed to hold quite a different view. Thus he wrote that

The meagreness with which attitudes are represented in consciousness resulted in a tendency to regard them as manifestations of brain activity or of the unconscious mind. The persistence of attitudes which are totally unconscious was demonstrated by Müller and Pilzecker (1900), who called the phenomenon 'perseveration'. (Allport 1935: 801)

So in this classic chapter, Allport not only displays explicit recognition of the significance of the unconscious dimensions of an attitude; he also praises Freud's contribution to this concept, and specifically applauds him for endowing attitudes with 'vitality, identifying them with longing, hatred and love, with passion and prejudice, in short, with the onrushing stream of unconscious life' (Allport 1935: 801). This all came from a man who had been personally put off by Freud's attempt to psychoanalyse him in his office some fifteen years previously.

But his chapter has a historical time line underpinning it: as we get towards the middle and end of the chapter the unconscious is mentioned less and less, and by the final section the focus has moved entirely from the unconscious to the conscious – in effect, to what can be measured with the greatest ease. He seems impressed by Likert's research which looked at white people's attitudes to 'Negros' and the fact that scales could be used to determine 'the amount of favor or disfavor toward the rights of the Negro'. The 'Likert scales' were measurement devices that pick up on conscious thoughts: sometimes thoughts that maybe we are not that happy with, but conscious thoughts nonetheless. Allport seems in awe of the work that had been done on intelligence testing, despite the fact that there was still huge disagreement about what 'intelligence' actually was. He wrote admiringly about the domain of intelligence, 'where practicable tests are an established fact although the nature of intelligence is still in dispute' (and, of course, he is alluding here to the huge debate then raging about how to define the concept of 'attitude').

These intelligence tests yielded vast amounts of quantitative data which in his view were clearly of practical value to an emergent nation. He saw the same practical application for attitude measurement and he concluded with some pride that 'The success achieved in the past ten years in the field of the measurement of attitudes may be regarded as one of the major accomplishments of social psychology in America. The rate of progress is so great that further achievements in the near future are inevitable' (Allport 1935: 832). As soon as he found himself in the domain of intelligence testing and Likert, the unconscious dimensions of the attitude seem to have been forgotten and remained forgotten for a significant period of time. Allport wanted to measure attitudes with precision and reliability, so he went for introspective paper and pencil-type tests. He clearly had his own objection to hypothesising scientifically about unknown and unknowable forces affecting our lives through the unconscious and, given Allport's influence on the developing field of social psychology, the focus turned away from the unconscious to the conscious, and therefore to self-report-type measures. It stayed there for the next half-century and more.

So, perhaps not surprisingly, I decided to begin my research programme proper by measuring attitudes towards carbon footprint in the tradition of Gordon Allport and the great experimentalist social psychologists who have documented our attitudes for decades, using Likert scales to reveal consciously held attitudes. It was a safe and obvious bet.

What did we already know about environmental attitudes from the available sources? The picture in the published literature is very positive (indeed, almost too positive): generally speaking people say they are extremely concerned about the environment and really do want to make a difference (although some are more than a little confused about what carbon labelling actually means). Consider the short summary of previous research in Table 3.1, which seems quite typical:

Table 3.1 **Explicit attitudes towards climate change and environmental behaviour**

Explicit measure source	Findings
70% of people agree that if there is no change in the world, we will soon experience a *major environmental crisis*	Downing and Ballantyne (2007)
78% of people say that they are *prepared to change* their behaviour to help limit climate change	Downing and Ballantyne (2007)
85% of consumers want *more information* about the environmental impacts of products they buy	Berry, Crossley and Jewell (2008)
84% say retailers should do more to *reduce the impact* of production and transportation of their products on climate change	Ipsos MORI (2008)
Only 38% of the general public *understand* what 'carbon labelling' means	Ipsos MORI (2008)

There seemed to be something of a clear consensus here, but what would my new research uncover? I used two measures – a computerised Likert scale and a 'feeling thermometer', which in this case would assess explicit feelings of warmth or coldness towards products with a high or low carbon footprint. The Likert scale gives a very simple measure of underlying attitude. A typical Likert scale is shown in Figure 3.1, and this is the computerised version we used in our actual research.

The feeling thermometer asked people to rate how warm or cold they felt towards high-carbon-footprint and low-carbon-footprint products and then the experimenter

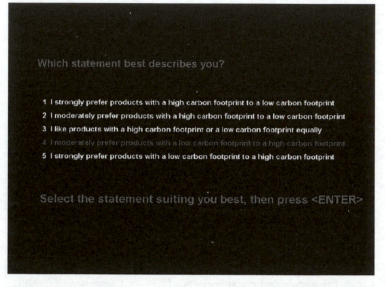

Which statement best describes you?

1 I strongly prefer products with a high carbon footprint to a low carbon footprint
2 I moderately prefer products with a high carbon footprint to a low carbon footprint
3 I like products with a high carbon footprint or a low carbon footprint equally
4 I moderately prefer products with a low carbon footprint to a high carbon footprint
5 I strongly prefer products with a low carbon footprint to a high carbon footprint

Select the statement suiting you best, then press <ENTER>

Figure 3.1 **A computerised version of the Likert scale for measuring attitudes to carbon footprint.**

computed the difference between these two numbers. For example, someone with a very positive attitude to low-carbon-footprint products might tick '5' meaning 'very warm' to the low-carbon-footprint products and '1' meaning 'very cold' to the high-carbon-footprint products, and this would yield a thermometer difference score of '+4'. On the other hand, someone who had a very positive attitude towards high-carbon-footprint products might tick '5' meaning 'very warm' on the high-carbon-footprint product and '1' on the low-carbon-footprint product, thus producing a thermometer difference score of '−4' (see Figure 3.2).

We found our first sample of participants for this research (Laura Sale was now officially my research assistant); we wanted to find a range of people of different ages and of different social backgrounds (not just the usual university students, but as usual somehow convenient for the environment of the university or college). Each participant was run individually and we had to do it this way to make sure that they knew what a carbon footprint actually was. In one

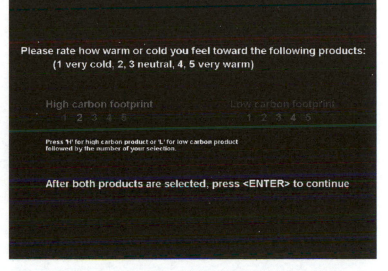

Figure 3.2 **A computerised version of the feeling thermometer scale for measuring attitude towards high- and low-carbon-footprint products.**

subsample of college kids, aged sixteen and upwards, we had to explain each time what a carbon footprint was before they could fill in the scale. Each time we administered the computerised test we had to check that they actually knew what they were rating. One seventeen-year-old lad interpreted the symbol quite literally (even after our introduction) and thought that it was the dent you made on the earth's crust as you went about your everyday business. I never actually met this person and I thought that he had perhaps just a very visual way of understanding the nature of the world and how it works. Others were less kind.

The results looked very promising. The Likert scale revealed that 30% of our participants demonstrated a preference for products with a low carbon footprint and 40% of participants demonstrated a moderate preference for products with a low carbon footprint; 26% of participants demonstrated no preference and only 4% demonstrated a preference for products with a high carbon footprint (see Figure 3.3).

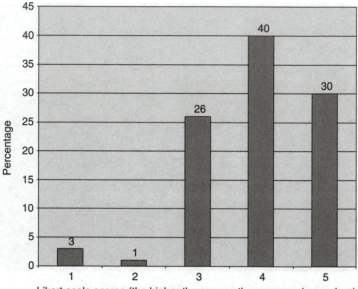

Figure 3.3 **Explicit attitudes to carbon footprint as revealed by the Likert scale.**

The feeling thermometer produced very similar results – 23% of participants showed a very strong preference for products with a low carbon footprint and an additional 40% showed some preference for low carbon products; 26% of participants were neutral in their attitudes (exactly the same figure as revealed by the Likert scale) and 7% of participants showed a preference for products with a high carbon footprint (see Figure 3.4).

These are clearly positive results, but just how positive depends on how you look at them. One way of looking at them (the way of the optimist) is to conclude that somewhere in the region of 67% to 70% of participants already show some significant preference for low-carbon-footprint products. The other way of looking at the data is that only 23% to 30% of people showed a strong preference for low-carbon-footprint products and that in reality we need to work on 70% of the population that are not sufficiently

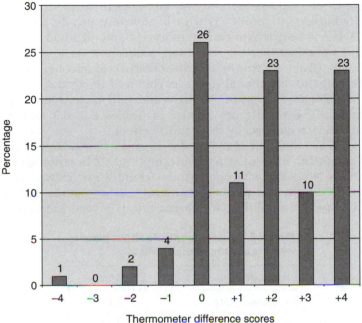

Figure 3.4 **Explicit attitudes to carbon footprint as revealed by the feeling thermometer.**

concerned about this issue. But these results would suggest that underlying attitudes are very good.

However, there is a slight fly in the ointment that should be obvious to everyone. What happens if the expression of these attitudes is affected by the fact that everybody in our society knows that green is good? Even the lad who thought that the carbon footprint of a product reflected indentations on the earth's crust still knew to say that low is good. How do we know that all of these easily expressed and deeply felt underlying attitudes are something more than the desire to look good in a research encounter?

We checked to see how people thought about those who cared for the environment and, not surprisingly, they thought in very positive terms about them. We found a new group of respondents and we asked them to describe the attributes of someone (1) who is environmentally friendly, (2) who is sensitive to carbon footprint and (3) who recycles, and we asked them to rate their judgement on a series of scales (from +3 to −3) – 'considerate and inconsiderate', 'thoughtful and thoughtless', 'caring and uncaring', 'knowledgeable and ignorant', 'selfless and selfish', and 'nice and nasty'. The results are shown in Tables 3.2, 3.3 and 3.4 and show a clear social desirability effect.

Who wouldn't want to be seen as 'considerate' and 'thoughtful' and 'knowledgeable' and 'nice'? In other words, people who are environmentally friendly are viewed in a positive light and the problem with this, of course, is that this social desirability, which is obviously very widespread

Table 3.2 Social judgements about people who are environmentally friendly

Attribute	Mean
• Considerate	1.68
• Thoughtful	1.64
• Caring	1.27
• Knowledgeable	1.14
• Selfless	1.09
• Nice	0.64

Table 3.3 Social judgements about people who are sensitive to a carbon footprint

Attribute	Mean
• Thoughtful	1.41
• Considerate	1.36
• Knowledgeable	1.32
• Caring	1.27
• Nice	0.73
• Selfless	0.68

Table 3.4 Social judgements about people who recycle.

Attribute	Mean
• Considerate	1.64
• Thoughtful	1.59
• Knowledgeable	1.45
• Caring	1.18
• Selfless	0.91
• Nice	0.59

throughout society, could potentially affect the expression of explicit attitudes towards environmental behaviour. In situations like this, it is clearly too risky to rest all conclusions on what people tell us either in interviews or even in the relevant anonymity of a computerised scale. People know that they are being judged: when you need to do research on a one-to-one basis there is always some kind of social bond between the researcher and the researched, and most human beings want to come across well. For these reasons and more I began to worry about how we get at underlying attitudes in situations like this. And then, of course, there was a more basic worry at the back of my mind – perhaps, after all, people do not know what their underlying attitude really is; perhaps Allport was wrong in the way that he developed the concept of attitude. Perhaps it does have its real roots in the unconscious. Perhaps he was right the first time when he wrote that 'often an attitude seemed to have no representation in consciousness other than a vague sense of need, or some indefinite or unanalyzable feeling of doubt, assent, conviction, effort, or familiarity' (Allport 1935: 800). Perhaps this was what we had to try to measure. It turned out that I was not alone in my doubt; someone had got there long before me.

4

The man who changed a fortune cookie and started a revolution

If attitudes were measured easily and uniformly for the first sixty years of their existence, in the 1990s there was something of a revolution in the measurement of attitudes and indeed in our whole approach to the topic. What drove this revolution was what psychologists like to call the 'attitude–behaviour problem' – the fact that attitudes and behaviour often simply fail to match up, and that attitudes measured in this way often do not predict behaviour in quite the way that many had hoped.

In an article in 1990 Anthony Greenwald attacked the then current theoretical understanding of attitudes and developed an argument that we should reconceptualise what we mean by an attitude in the light of new research in cognitive rather than social psychological research (the principal domain for this type of work). He cited the conclusion of Myers (1987), who had come to the view that the models of the attitude–behaviour relationship only really worked by 'limiting the scope of the attitude concept'. Thus:

> Our attitudes predict our actions (1) if other influences are minimized, (2) if the attitude is specific to the action, and (3) if, as we act, we are conscious of our attitudes, either because something reminds us of them or because we acquired them in a manner that makes them strong. When these conditions are not met, our attitudes seem disconnected from our actions. (Myers 1987: 45)

Greenwald, clearly not a man to mince his words, wrote:

'Myers' conclusion is decidedly embarrassing as a summary of the predictive power of social psychology's major theoretical construct' (1990: 256). What he did in the remainder of this short paper was to review new research in cognitive psychology on unconscious cognitive processes to provide a new theoretical basis to the work on attitudes. The fully formed Gordon Allport, the psychologist who had rejected Freud's attempt to bring the role of the unconscious even into his brief meeting with his younger self, would have turned in his grave. One consequence of Greenwald's devastating review was to 'call into question the appropriateness of the presently most favoured techniques of attitude measurement' (1990: 256). The rest, as they say, is history.

So who was this Anthony Greenwald? Greenwald was an American professor who had obtained his PhD in Social Psychology from Harvard in 1963, interestingly under the direct tutelage of Gordon Allport. As something of a rising star, Greenwald became known for his ability to take old concepts and theories and breathe new life into them through his singleminded determination (according to his colleagues). Mahzarin Banaji, who has worked with him for many years, illustrated Greenwald's eye for improvement with a telling anecdote during her congratulatory speech for Greenwald when he was honoured with the 'Recipient of the Distinguished Scientist Award, 2006'. Banaji recalled, 'I was there when you opened that famous fortune cookie message. It read innocuously, "There is nothing which cannot be improved." Not good enough for you!' Greenwald carefully reached for his pen, and drew a line through the word 'which', replacing it with the word 'that', so the fortune cookie now read, 'There is nothing *that* cannot be improved.' It takes a certain chutzpah to correct a fortune cookie in front of your colleagues, especially to correct the grammar of a fortune cookie in public. Greenwald was to demonstrate this chutzpah repeatedly in the development of this new field, a field that is so closely linked to this day with the name of this one eminent researcher.

Greenwald started with Myers' conclusion. Myers' paper was in many people's eyes a devastating summary of the attitude–behaviour relationship, but one bit of the conclusion was wrong according to Greenwald, and that was his third conclusion. Greenwald claimed that some of the most reliable and robust findings of attitudes predicting behaviour are exactly in those domains where the actor is not attentionally focusing on the attitude. One example of this, used by Greenwald, is the halo effect. The halo effect is the tendency to make new positive (or negative) judgements about a person when a positive (or negative) attitude towards that person already exists. In a famous study Landy and Sigall (1974) found that male participants judged the quality of a poor essay more favourably when the female author was attractive than when she was unattractive. This seemed to occur without any attentional focus on the underlying attitude. The fact that a photograph of the author was in the folder with the essay was designed to be almost coincidental.

Greenwald did not attempt to excuse himself here. He used an example from his own life as an academic to illustrate that implicit attitudinal forces are in constant operation in our everyday lives:

As a manuscript reviewer, I often cannot help noticing an initial warm, positive reaction when I review a manuscript that cites my work favourably (or maybe just cites it at all), and sometimes I notice the opposite – a colder reaction when some of my work that might have been cited is not mentioned. I know that these reactions interfere with the way my work as reviewer should be done, but it is difficult to avoid these reactions – and it is difficult not to do the review by searching for virtues that will justify the initial warm reaction, or for flaws that will justify the initial cold reaction. (Greenwald 1990: 257)

This was the phenomenon that he was trying to pin down and understand – this initial reaction full of emotional overtones well under the radar of consciousness but ultimately controlling our behaviour. Nobody, but nobody, escaped his gaze. With Eric Schuh and Katharine Engnell he analysed the citation patterns for authors whose names were selected from the 1987 Social Sciences Citation Index on the basis that they could be classified unambiguously as Jewish or Anglo-Saxon in origin. They found that Jewish-named authors cited 6% more authors with Jewish names than did Anglo-Saxon authors, and conversely Anglo-Saxon authors cited 7% more authors with Anglo-Saxon names than did Jewish authors (Greenwald, Schuh and Engell 1990) His conclusion was that 'Social scientists (who are widely regarded as being relatively free of prejudice) might display ethnic prejudices' (Greenwald 1990: 258). It seemed that even his friends and colleagues were not safe from his emerging theoretical views. This was putting colleagues on the spot just as pointedly as did Freud exactly seventy years earlier.

Greenwald saw an emerging pattern in all of this.

> The subject is in a situation that requires a response to some object; attitude towards an attribute of the object influences the response, and it does so without the subject's being aware that an attitude is being activated. These situations amount to indirect memory tasks for which the response has an evaluative component.
> (1990: 259)

Greenwald's goal was to elucidate the processes underlying implicit cognition and to demonstrate why they were critical to a reformulation of how we thought about attitudes and how we should measure them.

For Greenwald and Banaji in 1995, 'The signature of implicit cognition is that traces of past experience affect some performance, even though the influential earlier experience is not remembered in the usual sense – that is, it is unavailable to self-report or introspection' (1995: 4–5). This paper provides a simple example of implicit cognition in operation. Consider an experiment where the participants have to generate a complete word in response to an in-complete letter string (a word stem or a word fragment). The words that the participants generate here are more likely to be words that they have been 'casually' exposed to earlier in the experiment than words that they have not been exposed to. Greenwald and Banaji write that 'This effect of prior exposure occurs despite subjects' poor ability to recall or recognize words from the earlier list' (1995: 5). In other words, even though the participants do not recognise certain words as being on a list that they have seen previously, these words have been 'primed' in their memory and come out more readily than words that have not been primed.

Similar to this is the experimental work he had already conducted on what he called 'detectionless processing', one of the main areas of implicit cognition, in which stimuli of which people have little conscious awareness can be demonstrated to have an impact on behaviour. In detectionless processing, he demonstrated that words can be processed in terms of meaning, and therefore have an impact on our subsequent thoughts, even in situations where the words are presented below the threshold of conscious perception. What the participants in his study had to do was to decide whether each of a series of target words meant something good (e.g. 'fame', 'comedy', 'rescue') or bad ('stress', 'detest', 'malaria'). Half a second before each of these target words was presented, a priming word was presented briefly to the non-dominant eye. The priming words themselves were either good ('happy', 'joy', 'peace', 'love', 'excellent', 'pleasant') or bad ('evil', 'grief', 'sad', 'gloom', 'ugly', 'horrid'). The priming word was followed a matter of milliseconds later by a pattern mask (a complex visual stimulus) that interferes with the perception of the word that precedes it such that participants were unable to report what position the priming words were in on a computer screen (i.e. whether they were to the left or right of a fixation point). So even though the participants seemed to have little conscious information about the priming words and did not know even where these words had been presented, these 'invisible' words affected processing of the subsequent target words, such that target words that were preceded by a congruent prime (for example, a positive target word preceded by a positive prime, or a negative target word preceded by a negative prime) were identified significantly more quickly than target words that were preceded by a incongruent prime (see Greenwald, Klinger and Liu 1989). In other words 'invisible' words (with extremely brief presentations followed by a pattern mask) can influence the workings of the human mind. Suddenly, the unconscious was back in vogue.

The article that Greenwald wrote with Mahzarin Banaji in 1995 is in many senses of the word a classic paper. It has been cited by social scientists over 1200 times. The thesis of the paper articulated Greenwald's view of the unconscious in determining our behaviour. It successfully links the growing body of research on implicit cognition with the research on attitudes and behaviour. It argued that 'Recent work has established that attitudes are activated outside of conscious attention, by showing both that activation occurs more rapidly than can be mediated by conscious activity ... and that activation is initiated by (subliminal) stimuli, the presence of which is unreportable' (1995: 5).

Greenwald was keen to demonstrate the role of implicit processes in a series of domains. He employed the 'false-fame procedure' in an attempt to find experimental evidence for implicit stereotyping, including sex-role stereotyping which associates gender with achievement. The method is simple yet produces highly significant results. It was based on a procedure used by Kelley and his colleagues (Jacoby, Kelley, Brown and Jasechko 1989) to uncover implicit memory. In the Kelley studies, participants would read a list of both famous and non-famous names on day 1. The next day the same participants were presented with previously seen non-famous names from the first list or new non-famous names that they had not seen before mixed in with previously seen and new famous names. The question that they were asked was 'Is this person famous?' The researchers hypothesised that although the non-famous names should fade from memory over the twenty-four-hour period (from day 1 to day 2), the fact that the participants have seen some of the non-famous names before on the first list might lead them to the conclusion that some of these names were actually famous. And that is exactly what they found: there was a higher false-alarm rate for the previously seen non-famous names than for the new ones. Some non-famous names had quite literally 'become famous overnight' because of the extra exposure and the resulting familiarity. This was a demonstration of a significant unconscious influence on memory.

What Greenwald and Banaji did next was to introduce gender into this simple paradigm. They asked what happens if you use a set of male and female names here and consider gender as a principal variable. They found that the false-alarm rate for (previously seen) non-famous names was significantly greater for male than for female names. In other words, the participants were more likely to assume that the male names were famous (because of the familiarity due to repeated exposure) when they were not actually famous, compared with female names. So the authors concluded that this was clear evidence for implicit gender stereotyping in everyday life where maleness is associated with achievement.

In their summing up in this article, they concluded that:

considerable evidence now supports the view that social behaviour often operates in an implicit or unconscious fashion. The identifying feature of implicit cognition is that past experience influences judgment in a fashion not introspectively known by the actor. The present conclusion – that attitudes, self-esteem, and stereotypes have important implicit modes of operation – extends both the construct validity and predictive usefulness of these major theoretical constructs of social psychology. Methodologically, this review calls for increased use of indirect measures – which are imperative in studies of social cognition. (1995: 4)

In other words, this review article was arguing for a major reconceptualisation of how we should view both attitudes and social behaviour more generally. And, of course, if we accept the argument about the role of implicit or unconscious factors in controlling both our attitudes and our behaviour then we must accept that we will need a new methodology for studying some of these processes: a new methodology that does not involve self-report and is sufficiently sensitive to detect some of the very rapid processes involved. The development of the new method was crucial here because Greenwald recognised that some of what he was saying had been said before.

Implicit social cognition overlaps with several concepts that were significant in works of previous generations of psychologists. Psychoanalytic theory's concept of cathexis contained some of the sense of implicit attitude, and its concept of ego defense similarly captured at least part of the present notion of implicit self-esteem . . . At a time when the influence of psychoanalytic theory in academic psychology was declining, its conceptions of unconscious phenomena that related to implicit social cognition were being imported into behaviour theory (Dollard and Miller 1950; Doob 1947; Osgood 1957). The New Look in Perception of the 1950s focused on several phenomena that are interpretable as implicit social cognition. The developing Cognitive approach to these phenomena can be seen in Bruner's (1957) introduction to the concept of perceptual readiness. (Greenwald and Banaji 1995: 20)

But the punch line came next. Here after all was the man with all the chutzpah to correct the grammar of a fortune cookie in public.

Importantly, the psychoanalytic, behaviourist, and cognitive treatments just mentioned all lacked an essential ingredient, that is, they lacked reliable laboratory models of their focal phenomena that could support efficient testing and development of theory. The missing ingredient is now available. (Greenwald and Banaji 1995: 20)

5

The missing ingredient is now available

The missing ingredient was introduced in 1998 in an article published in the *Journal of Personality and Social Psychology* and coauthored with Debbie McGhee and Jordan Schwartz. Greenwald introduced the new method with a thought experiment. Imagine being shown a series of male and female faces and having to respond rapidly by saying 'hello' to the male face and 'goodbye' to the female face. Then you are shown a series of names and this time you say 'hello' to the male name and 'goodbye' to the female name.

In that experiment subjects gave a response on a computer keyboard with the index finger of the right hand to words that named pleasant things and to names of flowers. With the left hand they were to respond to another two categories – words that named unpleasant things and insect names. This was a very easy task. Then we made one minor change: We switched hands for the flower and insect names. Now subjects had to give the same response to pleasant words and insect names and a different response to unpleasant words and flower names. Immediately the task became hugely difficult. The slowing on a response-by-response basis was on the order of 300 milliseconds, which was a magnitude of impact nobody could have expected. We certainly did not expect it.

I was the first subject in the experiment. When I experienced the slowing I found to my surprise that I could not overcome it – repeating the task did not make me faster. If I tried to go faster, I just started making errors when I was trying to give the same response to flower names and unpleasant words. This was a mind-opener.

The very first paper reporting the Implicit Association Test (IAT) provided psychologists with a much sought-after method to measure unconscious, implicit attitudes; but perhaps even more than that, it uncovered something that was extremely unsettling for Greenwald and colleagues, and no doubt for anyone who read the paper. In today's society we like to think that race is no longer a significant issue: I am writing this particular paragraph in the same week that America has just elected its first Black president; surely the times of racial prejudice and stereotype are far behind us all in the West. The IAT revealed that this is not necessarily the case. The basic premise behind the IAT is that when categorising items into two sets of paired concepts, if the paired concepts are strongly associated, then participants should be able to categorise items faster into these category concepts. The IAT revealed that people were consistently faster at categorising Black and White names and pleasant and unpleasant words when the target categories were grouped 'White'/'pleasant' and 'Black'/'unpleasant' than when they were grouped 'White'/'unpleasant' and 'Black'/'pleasant', suggesting that the former concepts are strongly associated. When compared to explicit measures, the majority of White college students who took part in the study reported that they had no racial preference between White and Black, with some even saying they had a preference for Black. However, the IAT revealed that only one of these students showed a preference for Black consistent with their stated explicit attitudes. The remaining participants all showed a White preference, suggesting that White had positive associations, whereas Black had negative associations. As such, the IAT was able to successfully reveal underlying implicit attitudes that firstly cannot be masked by social desirability concerns and secondly a person may be totally unaware of holding.

In the first web-based experiment of its kind, Project

Implicit measured implicit attitudes towards a range of social groups, including implicit measures of racial attitudes. The project collated a staggering 600,000 tests between October 1998 and December 2000, allowing for replication of the race IAT on an enormous scale using both White and Black participants, with surprising results. It found that White participants tended to *explicitly* endorse a preference for White but *implicitly* they demonstrated an even stronger preference for White names and faces. Black participants, on the other hand, demonstrated a strong *explicit* preference for Black yet remarkably in the IAT, Black participants demonstrated a weak *implicit* preference for *White* names and faces. According to Nosek and colleagues who provided this overview of the IAT results in 2002, the preference shown for White by both White and Black participants is indicative of the American culture in which Black Americans are still often depicted in a negative light. The result is that these negative associations have penetrated into underlying racial attitudes and stereotypes, leading to the creation of automatic evaluations which show an implicit preference for White over Black people. For people who show strong explicit endorsement of racial indifference, the prospect that they may implicitly hold the very attitudes they strongly condemn can be a worrying thought (see also Gladwell 2005). As Fyodor Dostoevsky (1864) wrote in his *Notes from the Underground*:

> Every man has some reminiscences which he would not tell to everyone, but only to his friends. He has others which he would not reveal even to his friends, but only to himself, and that in secret. But finally there are still others which a man is even afraid to tell himself, and every decent man has a considerable number of such things stored away. That is, one can say that the more decent he is, the greater the number of such things in his mind. (1864/1972: 55)

However, the research has also suggested that implicit attitudes can be changed. In 2001, Greenwald and Dasgupta found that by exposing participants to pictures of a range of admired Black Americans such as Martin Luther King and disliked White Americans such as Al Capone, the pro-White effect usually found in the race IAT was substantially reduced, immediately after and even twenty-four hours after the initial exposure. While this was only a temporary modification, there is the possibility that being consistently exposed to exemplars of admired Black people (particularly in the media) could lead to more permanent changes in underlying implicit attitudes.

It just so happened that Laura took the US Election 2008 IAT on Project Implicit in the week after Barack Obama was elected as the next US President. Her results speak for themselves: 'Your data suggests a strong automatic preference for Black people over White people' and 'Your data suggests a strong automatic preference for Barack Obama over John McCain'. This could be an example of the malleability of implicit attitudes operating in the real world; all the positive exposure to Obama during that week, in all probability, had a significant effect on Laura's implicit attitude. Continued positive exposures to Black role models could lead to more permanent positive associations for Black people in general.

At present there are something like fifteen versions of the IAT online at Project Implicit:

- Disability IAT
- Asian IAT
- Religion IAT
- Skin-tone IAT
- Weapons IAT
- Age IAT
- Arab–Muslim IAT
- Sexuality IAT
- Weight IAT
- Gender–Career IAT
- Gender–Science IAT
- Native IAT
- Obama–McCain IAT
- Presidents IAT
- Race IAT

For a measure of unconscious processing, engaging on the IAT is an oddly self-conscious process. I am strangely anxious every time I do it, maybe because I think that this may reveal the uncomfortable truths about me. It is a quick test, almost too quick, and the computerised IAT flashes the results up at you without embarrassment or pause after the completion of each test. You sit nervously by the screen prepared to view your own prejudices and biases, secretly hoping that none will be revealed. Or at worst, hoping to see just a slight prejudice in your reaction times and error rates, and the expression 'Your data suggest a moderate preference for X over Y.' Laura and I both sat all the tests that we thought might help produce a reasonable psychological profile for each of us, one after the other like a set of challenges. The results are given in Table 5.1.

Table 5.1 Our own IAT results

	Geoff	Laura
Obama–McCain IAT	Your data suggest a strong automatic preference for Barack Obama over John McCain	Your data suggest a strong automatic preference for Barack Obama over John McCain
Race IAT	Your data suggest a moderate automatic preference for European American over African American	Your data suggest a moderate automatic preference for European American over African American
Skin-tone IAT	Your data suggest a moderate automatic preference for light skin over dark skin	Your data suggest a strong automatic preference for light skin over dark skin
Weight IAT	Your data suggest a strong automatic preference for thin people compared to fat people	Your data suggest a moderate automatic preference for thin people compared to fat people
Sexuality IAT	Your data suggest a moderate automatic preference for straight compared to gay people	Your data suggest a moderate automatic preference for straight compared to gay people
Age IAT	Your data suggest no automatic preference for young compared to old.	Your data suggest a slight automatic preference for young compared to old.

We discovered that we were both strongly pro-Obama, which was fine, even a little reassuring (to my conscious mind). I admit that I could never take John McCain's voice seriously, because of its pitch and general tone and the fact that it sounded like something computer-generated by Disney, and I am sure that Mahmoud Ahmadinejad, Hassan Nasrallah and Muqtada al-Sadr couldn't take it too seriously either. There was not much threat in that voice and the danger would always have been that the voice would have had to be backed up with what the military, and the new head of Central Command General David Petraeus, were now calling 'kinetics' (a term interestingly borrowed from mainstream psychology but now being used to refer to military action rather than action in general). After that I either had no preference (age) or a series of moderate preferences, except when it came to weight, where I had a strong preference for thin people compared to fat people (Laura's only other strong preference was for light skin over dark skin).

So what does any of this mean? Can I find any evidence from my own life that the implicit attitudes revealed by this test have any substantive or actual behavioural implications? I think that the answer with respect to my one strong prejudice (other than Obama) is probably yes. However, the behaviour is not to do with actual discrimination against fat people but behaviour directed against myself, and not dieting but something else. I have always been a compulsive runner, and compulsive here means *compulsive*, every day without fail, in any country no matter how inconvenient or difficult: through the centre of Tokyo at 5.00 a.m. because the flight back to the UK was to leave early; along a motorway in Sweden in the middle of the night in a snowstorm without a clue as to which direction led back to the centre of Gothenburg; along the Pacific Coast highway in California, just off the plane and suffering from jet lag, with no pavement for protection and with wide gaudy red and yellow trucks almost brushing my legs.

I was a child with a chubby face, never fat, but I am sure that strangers might have thought that I was fat because of my face ('you had a face like the moon when you were a baby,' my mother used to say proudly; 'like the moon,' and she would smile broadly whenever she said it, as if the memory made her happy) and I like the way that running makes my face look lean. I refuse to go on television unless I have a run first (many television producers will vouch in frustration for this fact).

But my one compulsion runs deeper than mere misplaced vanity, more rooted I am sure in my unconscious mind. This compulsion started when I was at school. I would run every day and twice on a Tuesday and a Thursday. I started when I was thirteen. A lot happened that year – I broke my arm doing judo, which meant that many sports were for a long while out of bounds, and my father died unexpectedly of some heart-related condition. I never understood what he died of, it was never properly explained to me as a child, and I am not sure that my mother properly understood what had happened anyway, except that it happened during an operation. I suppose that this made the fear more intense, the fear of life being interrupted in a sudden and unexpected way. I might have a similar congenital weakness, so I decided to get fit, and run and run to make my heart stronger and stronger. I never stopped. Running, we all know, can be very addictive.

But I have another image from that one life-changing year as well. An image that has never faded or been diminished by time, an image that has been silent and never discussed until now. An image, nevertheless, that has haunted me. On the night of my father's funeral, after we had laid him to rest in Roselawn Cemetery with the wind lifting the dirty green carpet used to cover the wet clay grave, we were in my aunt's house for the sandwiches and tea because our own house (a two-up, two-down in North Belfast) wasn't big enough. Everyone was there, drinking quietly, the quiet, subdued sobbing made worse by the image of the coffin juddering down into its final position: everyone except my cousin Myrna, that is, who inexplicably had gone to work that day. Nobody had explained why. Life at the time seemed to be full of things that were never properly explained, at least to a thirteen-year-old boy. My cousin walked in right in the middle of the wake. She seemed to cling to the doorframe, not entering, just standing there, staring at us all; and I can picture her now, an image etched on my mind for ever, an emaciated grey ghost, already dead in the eyes and the mouth. I had heard, overheard, that she had got some kind of eating disorder, anorexia nervosa – 'slimmer's disease', my mother called it – but I hadn't seen her for months, as the slimmer's disease took hold. She avoided seeing relatives. But there she stood in the silence and the sadness, and everyone looked at her and nobody said a thing, as if she looked normal and healthy and was just late for the funeral. I think she walked slowly past us all into the kitchen to stand alone, the place where food is prepared and eaten, but not for her.

She died a couple of weeks later of pneumonia and was buried in the row opposite my father, which is handy from the point of view of people bringing flowers to either grave. Her mother, my Aunt May, a sweet, lovely woman with a giggly, girly voice, always said that what triggered the anorexia was a chance remark from a doctor at work during a routine medical examination, a remark that she was a little overweight. From that day on, my aunt always said, she never ate properly again. It sounds almost ridiculous that such a life-threatening disorder could be triggered in this way, but years later I supervised a postgraduate student who analysed the social construction of anorexia in the families of sufferers and the number of interviewees who pointed to a similar 'chance remark' as the cause of the whole thing was extraordinary.

Anorexia is a complex disorder with cultural, personality and biological factors all implicated in its ontogenesis, but human beings like to identify a single cause that they can pick out and say 'if only that hadn't happened . . .'. This single cause is usually something fairly random (so that any random family could potentially be affected) and external to the family (so that no blame could be attached to the family). The PE teacher who commented that Tracy was too fat to be any good at games, the boyfriend who said that Jane's bum was too big for her skinny jeans, the doctor who quipped that his patient could do with losing a little weight. It was always things like that. It reminded me of what Friedrich Nietzsche wrote: 'To trace something unknown back to something known is alleviating, soothing, gratifying, and gives us moreover a feeling of power. Danger, disquiet, anxiety attend the unknown, and the first instinct is to eliminate these distressing states. First principle: any explanation is better than none' (1871/1962: 62).

My family, and many other families, had found an explanation, one that stressed the power of the word, of the chance remark (and I suppose, by implication, the dangerous power of the carelessness of those in authority). Nobody ever disputed my aunt's account, and it became the true version of what had become of my beautiful cousin, and in that awful year of my life I sometimes think that my weight prejudices were probably laid down for ever.

Of course, this story may tell you why weight is an important issue for me, but why have I ended up with an anti-fat prejudice, why not an anti-thin prejudice? After all, it was not the fact that Myrna was a few pounds overweight (maybe more, maybe less) that killed her. Well, maybe it was the implicit message in the story, the implicit message being that if you are overweight then you can be killed by a chance remark, the unconscious message being that being fat makes you too sensitive to others' insensitivity, the unconscious theme being that being fat means that others can control your life, and even your death. My compulsive running may just reflect my unconscious desire to escape from my father's destiny, but it may also reflect this deep-seated desire to put myself out of harm's way from chance remarks (and thereby make myself less vulnerable in life). It may be core to my psychological make-up and mean that I have an implicit and unconscious bias against fat people, who have not made the effort to shield themselves in this way. Of course the fact that my implicit attitude actually does connect to some core behaviours in my everyday life, namely my determination to run, is very encouraging from the point of view of my current academic concerns. It is also, of course, more than a little depressing for me.

But for the psychologist wanting to save the planet, the big question is whether the IAT gives us more insight than the measures of explicit attitudes. The beauty of the IAT is that because it measures automatically activated associations, it is resistant (some have argued 'immune') to faking (see Greenwald, Poehlman, Uhlmann and Banaji 2009). Would it, therefore, help us to understand and predict actual behaviour more effectively? The explicit tests that we had carried out had revealed that explicit attitudes are very pro-environment, so why wasn't people's behaviour falling in line with this? Would the IAT reveal something quite different here? Would this help explain what was going on? The focus on behavioural prediction was, of course, inevitable from the inception of the IAT and 'plagued' the IAT very early on (in the eyes and words of some of the core researchers). Banaji (2001) commented that 'tolerance' was needed if this question was to be answered: 'Pushing fast and furiously to "show me what predicts" may be counterproductive. One first needs to understand the construct before asking what it may or may not predict' (2001: 132).

Greenwald et al. (2009) conducted a meta-analysis on the predictive validity of the IAT and concluded that, in general, when the IAT and explicit attitude measures are combined, they are better predictors of behaviour than either measure alone. However, when attitudes are 'socially sensitive', and where social desirability concerns are inevitably present (as in attitudes to race, age, gender or the environment), explicit measures are very poor predictors of behaviour and in these situations the IAT would appear to be a much better predictor of behaviour than explicit measures.

Recently, Greenwald along with other researchers has started to apply the IAT to consumer research since it has become increasingly apparent that consumer behaviour does not necessarily involve conscious and rational decision-making, but can also be influenced by all sorts of unconscious factors. You can often see people in the supermarket picking up products without even properly looking at them, let alone making complex decisions based on price or nutritional value or fat content. There is something at work here that is not based on conscious, reflexive, rational thought and slow decision-making. And, of course, the IAT could be particularly useful in the domain of green consumerism given the evident attitude–behaviour gap in the purchasing of green goods. If implicit attitudes were measured in this particular domain, then a less optimistic view of environmental thinking may well be revealed, which reflects all the other concerns that consumers have about things like price and convenience, as well as potentially their concerns regarding their lack of knowledge about carbon footprint, and their general anxiety as to whether one person's individual shopping behaviour can actually make a difference.

From the meta-analysis conducted by Greenwald et al. (2009), the overall conclusion would seem to be that the IAT does significantly predict behaviour (although the actual level of prediction can be modest at times). But as Gregg (2008) has commented, 'the behaviours documented are often quite specific, so it is striking that general implicit associations predict them at all. Moreover, the IAT outstrips self-report in forecasting instances of discrimination and prejudice. Hence, it offers some genuine diagnostic advantages' (2008: 765). There appear to be something like fifteen published studies that have used the IAT in the consumer domain, as shown in Table 5.2. This table has been adapted from the Greenwald et al. (2009) paper, but I have tried to make the conclusions as specific and as accurate as possible by quoting directly from the original studies.

The trend in these results appears to be fairly positive with respect to the predictive power of the IAT and particularly under certain types of conditions. When people are under any kind of time pressure (like shopping in a supermarket, particularly just before closing time), or when they are having to control their emotional state (like shopping in a supermarket after a hard, stressful day at work), or when they are focusing on the kinds of enjoyment that they might get from their food choices (like shopping in a supermarket after a hard day at work and looking forward to dinner), or when they are under the influence of alcohol, the IAT is a very good predictor of behaviour. Measures of implicit attitude seem to predict behaviour best when the person concerned is under any kind of mental or emotional or time pressure: when there is a lot going on, in other words. Explicit measures, on the other hand, are usually better

Table 5.2 **The IAT as a predictor of consumer behaviour**

Citation	Behavioural measure	Did the IAT successfully predict behaviour?
Brunel, Tietje and Greenwald (2004)	Study 1: Self-report of ownership and usage frequency of Mac and PC.	Both IAT and explicit attitude measure predicted ownership and usage.
Friese, Hofmann and Wänke (2008)	Study 1: Behavioural choice task between apples and chocolate where working memory capacity is reduced.	When processing resources are 'ample', explicit attitude measure is a better predictor of behaviour. When processing resources are 'taxed', 'behaviour appeared to be more strongly driven by impulsive processes as indicated by the increase in the implicit measure's predictive validity'.
Friese, Hofmann and Wänke (2008)	Study 2: Consumption of potato crisps after watching a film where emotions were either controlled (depleting 'self-regulatory strength') or not controlled.	'. . . when participants were depleted of their self-regulatory strength [by having to suppress their emotional response to a film], not only did the implicit measure gain considerable predictive power compared with the control condition but also the explicit measure was now unrelated to potato crisps consumption.'
Friese, Hofmann and Wänke (2008)	Study 3: Beer consumption after watching a film where emotions were either controlled (depleting 'self-regulatory strength') or not controlled.	'When resources were scarce [because participants had to suppress their emotional response to a film] the implicit measure predicted behaviour well and showed incremental validity over and above both explicit self-report measures at the same time.'

Friese, Wänke and Plessner (2006)	Brand choice between generic and branded products in experimental conditions either under time pressure (5 seconds to make their choice) or not under time pressure (unlimited time).	'Participants whose explicit and implicit preferences regarding generic food products and well-known food brands were incongruent were more likely to choose the implicitly preferred brand over the explicitly preferred one when choices were made under time pressure. The opposite was the case when they had ample time to make their choice.'
Gibson (2008)	Brand choice between Coke and Pepsi in experimental conditions when cognitive load was manipulated (by asking participants to remember an 8 digit number, or not).	'. . . choice in this high load condition was related to implicit attitudes, while choice in the low load condition was not.'
Hofmann and Friese (2008)	Candy consumption when participants had been drinking alcohol or not.	'Specifically, the predictive validity of implicit attitudes (as part of the impulsive system) was markedly increased for participants who had consumed alcohol as compared with sober participants.'
Hofmann, Rauch and Gawronski (2007)	Candy consumption after watching a film and being asked either to suppress emotions (depletion condition) or to 'let emotions flow' (control condition).	'. . . automatic candy attitudes showed a positive correlation to candy consumption in the depletion condition but not in the control condition. That is, candy consumption significantly increased as a function of automatic positivity toward the candy in the depletion condition but not in the control condition.'

(Continued overleaf)

Table 5.2 – continued

Citation	Behavioural measure	Did the IAT successfully predict behaviour?
Karpinski and Hilton (2001)	Study 2: Behavioural choice between apples and candy bars.	'... explicit attitudes and the IAT are independent ... explicit attitudes predicted behaviour but the IAT did not.' (Note: there was no time pressure/drain on cognitive resources etc. operating here. In this study participants 'were informed that they could choose only one of the objects [apple or candy bar] to eat or to take home with them.')
Karpinski and Steinman (2006)	Study 1: Brand choice between Coke and Pepsi.	Both IAT and explicit measures predicted choice of branded drink.
Karpinski, Steinman and Hilton (2005)	Voting intention in the 2000 US Presidential election and also brand choice between Coke and Pepsi.	'... explicit attitude measures were better predictors of *deliberative* behaviours than IAT scores' (emphasis added).
Maison, Greenwald and Bruin (2001)	Study 1: Self-reported drinking of juices or soda, and self-reported dieting.	'... significant correlation between the IAT and Ss' *self-reported* behaviour' (emphasis in original).
Maison, Greenwald and Bruin (2004)	Study 1: Self-reported consumption of yoghurt brands/eating at different fast food restaurants/con-sumption of Coke or Pepsi.	'A meta-analytic combination of the three studies showed that the use of IAT measures increased the prediction of behaviour relative to explicit attitude measures alone.'

Olson and Fazio (2004)	Study 3: Self-reported behaviour of apple and candy bar consumption.	IAT predicted behaviour, particularly a more personalized IAT. '. . . the personalized IAT correlated more strongly with explicit measures of liking, past eating behaviour, and behavioural intentions than did the traditional IAT.'
Scarabis, Florack and Gose-johann (2006)	Choice between chocolate and fruit.	The IAT was a good predictor of actual choice, '. . . people rely more on automatic preferences that are independent from higher-order appraisals when they focus on their affective responses [what enjoyment they might get from the food] than when they think about the advantages and disadvantages of choice options.'
Swanson, Rudman and Greenwald (2001)	Study 2: Self-reported smoking behaviour or vegetarianism/non-vegetarianism.	The IAT and explicit attitude measures did predict vegetarianism/non-vegetarianism but not smoking.
Vantomme, Geuens, De Houwer and De Pelsmacker (2005)	Self-reported purchase intentions for real and fictitious brands of green and environ-mentally unfriendly cleaning products.	'The IAT, but not the explicit difference score, differentiated between respondents intending to buy the real ecological all-purpose cleaner and those intending to buy the real traditional all-purpose cleaner.'

when there is no mental load or time pressure, when there is all the time in the world, thereby allowing the person to make slower, deliberate and reflective behavioural decisions. Something like supermarket shopping, however, is not, generally speaking, a slow, deliberate, reflective process for most people. It is fast and non-reflective and sometimes quite hectic. So, in a context like this, the IAT should be a much better predictor of consumer behaviour than a measure of explicit attitudes.

Only one study seems to have applied the IAT to actual green consumerism. Vantomme and colleagues in 2005 conducted an experiment that looked at implicit and explicit attitudes towards green cleaning products. They expected to find that implicit attitudes would reveal less positive results than the explicit attitude measures (because of the social desirability factors relating to green behaviour). In their first experiment they used fictitious cleaning products, introduced to participants in a 'learning phase' where participants were informed that one product was environmentally friendly while the other was harmful to the environment. They used fictitious products because of the dangers of brand image impacting on the results. What they found, however, was that in contrast to what they had hypothesised, implicit attitudes towards the fictitious green cleaning products were far more positive than explicit attitudes.

In a second experiment, real brands were used instead. This time, there was no difference in implicit and explicit attitudes towards green cleaning products. So the results from this study are a little inconclusive with respect to the underlying implicit attitudes towards green products that people might actually hold. But both sets of results went against the original hypothesis. This study, therefore, left our own empirical investigation into implicit attitudes wide open, which made the whole thing, of course, that much more exciting.

6

Uncovering implicit attitudes to carbon footprints

We created our own version of the IAT which compared the categories of high- and low-carbon-footprint products and the attributes 'good' and 'bad' and tested it on a random sample; a mixture of students and ordinary working people. To get an idea of the IAT procedure for yourself, try the following example. All you have to do is put the pictures and words that appear down the middle of Tables 6.1 and 6.2 into the categories that appear on the left hand side ('Low Carbon Footprint or Good') or the right hand side of the page ('High Carbon Footprint or Bad') as quickly as you can. In the normal IAT, items are assigned to categories on the left by pressing a key to the left of the keyboard (e.g. 'Z') or to categories on the right by pressing a key to the right of the keyboard (e.g. 'M'). However, for the examples included here you can just tap either the left-hand side of the page or the right-hand side of the page. As in the experiment itself, you must try to do this as quickly as you can.

For example, the first item in Table 6.1, 'Awful', fits into the category 'High Carbon Footprint or Bad' because 'Awful' is clearly 'Bad', so tap the right-hand side of the page. (The underlying psychological reasoning here is that if you unconsciously think that high-carbon-footprint products are bad then this assignment should be relatively easy.) The second item, a Waitrose plastic bag, is clearly 'High Carbon Footprint' so here you should again tap the right-hand side of the page ('High Carbon Footprint or Bad'). Again, the reasoning is that if you unconsciously think that high-carbon-footprint products are bad then this assignment should be

Table 6.1 Sample IAT procedure 1

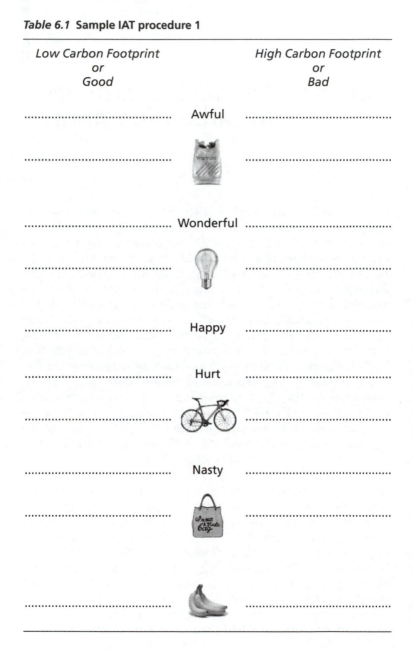

Low Carbon Footprint or Good		High Carbon Footprint or Bad
....	Awful	....
....		....
....	Wonderful	....
....		....
....	Happy	....
....	Hurt	....
....		....
....	Nasty	....
....		....
....		....

Table 6.2 **Sample IAT procedure 2**

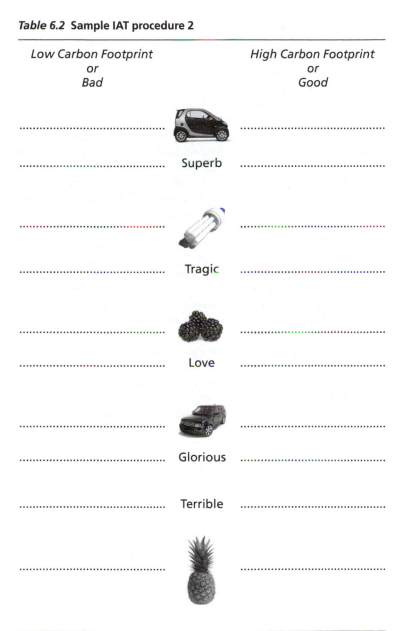

Low Carbon Footprint or Bad		High Carbon Footprint or Good
....		....
....	Superb	....
....		....
....	Tragic	....
....		....
....	Love	....
....		....
....	Glorious	....
....	Terrible	....
....		....

relatively easy. 'Wonderful' is 'Good' so now you should tap the left-hand side of the page, assigning 'Wonderful' to the 'Low Carbon Footprint or Good' category, and so on.

In Table 6.2, the category pairs have been reversed and we have 'Low Carbon Footprint or Bad' on the left-hand side of the table versus 'High Carbon Footprint or Good' on the right-hand side. The first item is a small eco-friendly car, so you should tap the left-hand side of the page assigning it to the 'Low Carbon Footprint or Bad' category because this small car is clearly 'Low Carbon Footprint'. The psychological reasoning here is that if your underlying unconscious attitude is very pro-low-carbon-footprint products then this will be (relatively speaking) harder to do than the previous task when the categories were paired differently, because the generic category you are assigning it to ('Low Carbon Footprint or Bad') also covers words or concepts that are 'bad'. The second item is 'Superb' so you should tap the right-hand side of the page ('High Carbon Footprint or Good') because 'Superb' clearly is 'Good'. Again this should be more difficult than before, if you unconsciously do not think that high carbon products are 'good'. The third item is a low-energy light bulb so you should tap the left-hand side of the page ('Low Carbon Footprint or Bad'), which is relatively speaking (and again we are talking about milliseconds here) more difficult for most people than the earlier task when the categories were paired in the reverse manner. So go ahead and try it for yourself to work out your own implicit biases.

If you do (in terms of your unconscious attitude) associate low carbon with 'good' and high carbon with 'bad' then you should have found the first table easier to do than the second. On the second table you may well have noticed a slowing in your reaction time. If, on the other hand, you associate low carbon with 'bad' and high carbon with 'good' (perhaps because you feel that people are trying to force green issues down your throat) then you should have been faster when categorising items in the second table, and your responses should have been slower when you were categorising items in the first table ('I was only getting started,' I hear you say; the mind is after all great at rationalising!).

Block	No. of trials	Items assigned to left-key response (z key)	Items assigned to right-key response (m key)
B1	20	'Low Carbon Footprint'	'High Carbon Footprint'
B2	20	'Good'	'Bad'
B3	20	'Good + High Carbon Footprint'	'Bad + Low Carbon Footprint'
B4	40	'Good + High Carbon Footprint'	'Bad + Low Carbon Footprint'
B5	40	'High Carbon Footprint'	'Low Carbon Footprint'
B6	20	'Good + Low Carbon Footprint'	'Bad + High Carbon Footprint'
B7	40	'Good + Low Carbon Footprint'	'Bad + High Carbon Footprint'

In Table 6.3, the blocks of trials used in the IAT (with the critical comparison trials outlined in bold) are shown:

The computerised versions of the seven trials are shown in Figures 6.1–6.7. This is what the participants actually saw on the computer screen in our IAT.

The D score (or difference score) is the critical measure used in the IAT. This is a statistical measure that calculates both the difference in the latency, or time taken to respond, in the critical trials *and* the error rate. The main point to remember here is: the more positive the D score, the more positive the implicit attitude to low-carbon-footprint products. The actual D scores we found are given in Table 6.4.

The results of our carbon footprint IAT revealed that 59% of our participants showed a strong implicit preference for products with a low carbon footprint. An additional 24% showed an implicit bias towards products with a low carbon

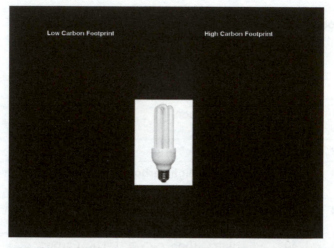

Figure 6.1 First trial: 'Low Carbon Footprint' vs 'High Carbon Footprint'.

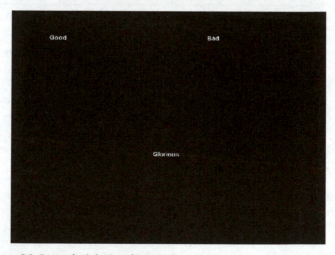

Figure 6.2 Second trial: 'Good' vs 'Bad'.

footprint. 10% of people were neutral, showing little or no preference for either high- or low-carbon-footprint products, and 7% showed an implicit preference for products with a high carbon footprint, as shown in Figure 6.8.

Put simply, implicit attitudes would seem to be even

Figures 6.3 and 6.4 Third and fourth trials:
'Good or High Carbon Footprint' vs 'Bad or Low Carbon Footprint'.

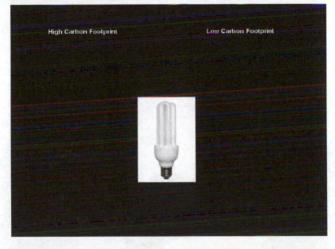

Figure 6.5 Fifth trial: 'High Carbon Footprint' vs 'Low Carbon Footprint'.

Figures 6.6 and 6.7 Sixth and seventh trial:
'Good or Low Carbon Footprint' vs 'Bad or High Carbon Footprint'.

Table 6.4 **D scores from the carbon footprint IAT**

D Score	Type of preference	Percentage
+0.8	Strong preference for low carbon	59%
+0.5	Medium preference for low carbon	15%
+0.2	Slight preference for low carbon	9%
0	No preference	10%
−0.2	Slight preference for high carbon	4%
−0.5	Medium preference for high carbon	3%
−0.8	Strong preference for high carbon	0%

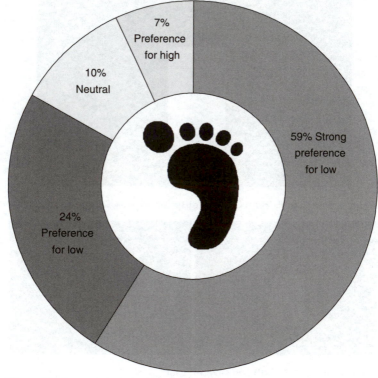

Figure 6.8 **D score percentages.**

more biased towards low-carbon-footprint products than explicit attitudes. Overall 83% of participants showed some preference for low-carbon-footprint products compared to 70% or 67% on the earlier explicit measures. While our explicit measures suggested that 26% of our participants held neutral attitudes, the IAT measure suggested that only 10% of participants held a neutral view towards high- and low-carbon-footprint products.

What could possibly account for the decrease in the proportion of people holding neutral attitudes? I have one possible 'methodological' explanation for what is going on here. It could be due to participants showing positive implicit bias towards *specific* products and images whereas, of course, on the explicit measures, participants are forced to imagine an undefined set of products with either high or low carbon footprints and therefore participants may have sat on the fence more with these abstract concepts. It is a possible explanation.

Of course the IAT measures implicit attitudes to high- and low-carbon-footprint products *generally*, but one can also focus on the response times to the different types of product used in the experiment and this does highlight some interesting differences between the products. By breaking down the overall response times into the individual mean response times for each picture item used in the IAT, we can gain an insight into how quickly our experimental participants categorised each individual item. This new focus reveals that certain food items, particularly fruit, are readily classified as having a high or low carbon footprint, suggesting that these items are easily automatically recognised as having either good or bad environmental characteristics. Pineapples (exotic, have to be transported great distances, therefore high carbon footprint, and therefore bad) were categorised most quickly of all, followed by English apples (low carbon footprint) and locally grown blackberries (low carbon footprint). The same is true for items representing certain energy sources (like wind) and modes of transportation (like cars). Oddly and somewhat counter-intuitively, the bicycle (the counterpoint to the car) took a lot of time to categorise.

It is only really within the past few years that packaging and the bags we use have been highlighted as a major environmental issue and an area where consumers can make a real difference, and the results showed that reusable carrier bags were the slowest to be categorised in this experiment. This result might surprise many people since the reusable bag has become such a symbol for whole sections of society who wish to flaunt their social identity as primarily concerned with green issues. It may act as a potent (and conscious and deliberate) social signal but it does not seem to have the same automatic, unconscious impact that some people might imagine. The results would seem to indicate that people need a little bit more time to process these iconic representations and assign them to one of two categories.

Similarly, our participants were quick to recognise that certain kinds of light bulb were good, but they took much longer to recognise that the alternatives (normal light bulbs) were bad. They also needed quite a lot of time to think about beef and chicken (the beef shown was high carbon footprint because cows generally have a higher carbon footprint than chicken and much beef comes from overseas). Figure 6.9 shows the response times for each IAT item.

So what are the implications of all this? The implications would seem to be that people have the right attitude, both implicit and explicit (both conscious and unconscious) to low-carbon-footprint products. Since such attitudes, and their combination, set up a predisposition to act, then one might expect people to be doing much more for the environment in terms of their everyday supermarket shopping than they actually are. Green choices are becoming more popular, but not as quickly as some might have imagined. So there is clearly something else going on here, but what sorts of additional processes are critical here?

And something else is evident in these data, the first clear hint that people might say one thing but believe another. The research had shown that people, generally speaking, were very pro-low carbon in both their explicit and implicit attitudes, but a significant proportion of people showed a marked discrepancy between these two measures. They were

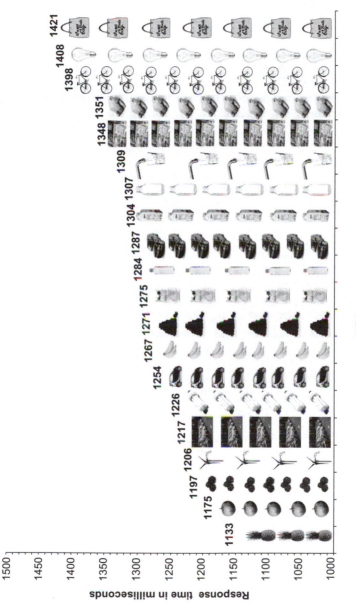

Figure 6.9 Response times to high- and low-carbon-footprint IAT items.

much more pro-low carbon in their explicit measure than in their implicit measure. This was the first hint in my data that some people might like to exaggerate their green credentials. They would report, when asked, that they were pro-low carbon products, and that, of course, they cared about the environment. The IAT, however, revealed something different. I was surprised to find (for all kinds of reasons) that I was actually one of them.

7

Unconscious eye movements and what the brain sees

I need to change tack now, to show the results of a very different sort of psychology experiment, one that was extremely time-consuming and laborious to carry out, as we now started to monitor the unconscious eye movements of individuals looking at packaging. (Luckily I had a second dedicated research assistant, Laura McGuire, to take the lead on this.) Why weren't people buying the low-carbon-footprint products given how positive their underlying attitudes were? I needed to track unconscious eye movements to determine what the brain actually sees when it looks at a product, to see whether this might hold a clue (perhaps they simply never noticed the carbon footprint information). This could prove to be illuminating in many ways. Thousands of minute dots on a computer screen had to be individually coded, but it was essential to finding out what is going on at the most basic level of human perception when people glance at products.

So how does this fit in with the bigger issues? Let me remind you. According to the Stern Review (2006: 1), 'The scientific evidence is now overwhelming: climate change presents very serious global risks, and it demands an urgent global response.' Many people have argued that the retail sector has a crucial role to play in this global fight against climate change (and, of course, it was this and similar arguments that persuaded Tesco to introduce carbon labelling on some of its products and to fund research in this area). According to Forum for the Future (2007):

Retail has a vital role to play in delivering sustainable development. It employs 2.9 million people and generates almost 6% of the GDP of the UK. It is responsible for approximately 2.5% of the UK's carbon dioxide emissions and has a disproportionate influence over society and the economy through its marketing, regular customer transactions and complex, globalised supply chains. (2007: 8)

Therefore, in order to reduce greenhouse gas emissions, it is vital that consumers reduce the carbon footprints of the products they buy. One common argument is that this is best done not through legislation or prohibition, by restricting what consumers can or cannot buy, but by providing all consumers with appropriate information about the carbon footprint of a product so that they can make an informed choice, effectively empowering consumers and revolutionising our patterns of consumption.

As Sir Terry Leahy (Chief Executive of Tesco) commented (2007), 'To achieve a mass movement in green consumption we must empower everyone – not just the enlightened or the affluent.' This philosophy resulted in the inclusion of carbon footprint information on an increasing number of products, which should guide consumers (*en masse*) to greener choices: assuming, of course, that they have the right underlying attitude to 'green' or low-carbon-footprint products in the first place. This latter point is crucial because if consumers don't have the right positive attitude to low-carbon-footprint products, it is unlikely that the inclusion of carbon footprint information will have any effect on consumer choice or behaviour. However, the studies that I have just outlined provide evidence that, in a sample of UK consumers in 2008, the vast majority not only had positive *explicit* attitudes to low-carbon-footprint products, as measured by the usual self-report measures, but also had very positive *implicit* attitudes to such products, as measured by the Implicit Association Test (IAT). These studies, therefore, give some psychological weight to the fundamental argument that we may, in fact, be able to reduce greenhouse gas emissions through consumer behaviour, because it demonstrates that many consumers (and not just a green minority) have exactly the right underlying attitude to low-carbon-footprint products in the first place. Therefore, there is a cogent argument that, in the case of many consumers, the provision of the carbon footprint information will allow their underlying attitude to be realised in actual behaviour.

But, of course, things might not be quite as simple as this. In everyday shopping, consumers are bombarded with product information and there is the danger, therefore, that the carbon footprint information will be not attended to, neglected or lost during the shopping experience. The carbon footprint information may be out there (on some products), but is it attended to and processed in the appropriate time frame? As Louw and Kimber (2007: 6) have noted:

> In a standard supermarket the typical shopper passes about 300 brands per minute (Rundh 2005). This translates into less than one-tenth of a second for a single product to get the attention of the customer and spark purchase (Gelperowic and Beharrel 1994: 7) . . . Even in high involvement situations, most consumers don't have the time, ability or information to assess all the pros and cons before purchase. Instead they rely on various cues (e.g. brand name, packaging etc) to help them make their decision. (Zeithaml 1988)

Carbon footprint information has to stand out in this general cognitive environment in which time is very much of the essence, and much information has to be ignored or processed very superficially before the purchasing decision is made.

So how is carbon footprint information currently represented on products? The form of representation currently being tested in the UK uses iconography (an image in black-and-white of an actual 'literal' footprint), accompanied by substantial amounts of text, numbers and scientific abbreviations both on the footprint (e.g. '12 kg CO_2', 'Compared to 100 W conventional 55 kg') and above and below the footprint (e.g. 'working with the Carbon Trust', 'per 1000 hrs of use', 'The carbon footprint of this light bulb is 12 kg per 1000 hours of use and we have committed to reduce the footprint of future equivalent light bulbs. By comparison the footprint for the equivalent conventional light bulb (100 W) is 55 kg per 1000 hours of use'). But all this information has to compete with large amounts of other information on the packaging (some also connected to 'green' issues, like the temperature that the detergent can be used at) on the fronts, backs and sides of products. For example, in three common products (sold in Tesco, UK), already labelled with carbon footprint information, this information has to compete with the information shown in Tables 7.1, 7.2 and 7.3.

These products thus have a range of pictures, diagrams, icons, numbers and text (with the text in different-sized fonts), all competing for the consumers' attention in a very

Table 7.1 **Information displayed on Tesco's low-energy light bulb packaging**

Light bulb – front view (in descending order of surface area)	Light bulb – back view (in descending order of surface area)
Product image (picture)	Carbon footprint (icon with text, numbers, scientific abbreviations)
'Greener living' (icon and phrase)	Carbon footprint information (text, numbers, scientific abbreviations)
'Bayonet cap' (diagram and name)	'Greener living' (icon and phrase)
'EDF energy' (icon and name)	'Bayonet cap' (diagram and name)
Wattage (number)	'EDF energy' (icon and name)
Life of the bulb (number)	Wattage (number)
Wattage equivalent (number)	Wattage equivalent (number)
Product name (bulb)	Product name (bulb)
Other	Other

Table 7.2 Information displayed on Tesco's freshly squeezed orange juice carton

Orange juice – front view (in descending order of surface area)	Orange juice – side view (in descending order of surface area)
Product image (picture)	Background image (picture)
Nutritional information (numbers, words and symbols)	Carbon footprint (icon with text, numbers, scientific abbreviations)
Product name and information (Tesco orange 100% pure squeezed juice)	Carbon footprint information (text, numbers, scientific abbreviations)
Price (text, numbers and icon)	'Picked and processed within 24 hours' (quote)
'NOT FROM CONCENTRATE' (quote)	Small product image (picture)
'With bits' (quote)	Price (text, numbers and icon)
Other	Other

limited time frame. Again, as Louw and Kimber (2007: 14) have noted:

> Shoppers typically only look at a label for about five to seven seconds, regardless of how many elements or messages there are on the package. Therefore, adding additional messages to the package increases the likelihood that a shopper will miss any single message. For this reason it is generally recommended that only two to three key points of communication are placed on a front label. Adding more messages is likely to clutter the label (which often detracts from appeal and perceived quality), and makes it more difficult for people to absorb the key information/communication from the label. (Young 2003)

This might be a key recommendation from some researchers in this area, but it does not prevent some retailers from

Table 7.3 **Information displayed on Tesco's 'Non-Bio' liquid detergent container**

Detergent – front view (in descending order of surface area)	Detergent – back view (in descending order of surface area)
30° (text and number)	Product instructions (text)
Product name (TESCO Non-Bio liquid detergent)	Product information (text)
Baby (picture)	Carbon footprint information (text, numbers, scientific abbreviations)
'Dermatologically tested' (quote)	Barcode (code)
'Suitable for sensitive skin' (quote)	Carbon footprint (icon with text, numbers, scientific abbreviations)
20 washes (number and text)	1.5 litres (number and text)
T-shirt (icon and text)	Other
Other	

including nine sets of key information on the front of the packaging of a light bulb (the one used in this experiment), seven on the front of the orange juice and eight on the front of the detergent. Moreover, the carbon footprint icon or the accompanying textual information is not represented on the front of any of these three products – instead it is represented on the backs of both the bulb and the detergent and on the side of the orange juice. The question then becomes: how salient is this carbon footprint information, especially under the time constraints of normal shopping, when many consumers are choosing products without much prior consideration (Hausman 2000; Silayoi and Speece 2004; see also Louw and Kimber 2007)? Time pressure must (by definition) reduce the detailed consideration of elements on a package, but how is attentional resource distributed to carbon footprint information compared to each of the other elements represented on the package?

One possible way of investigating this is by using eye-tracking technology to measure visual attention to each element on the packaging by tracking the overt movement of the eyes and measuring each period of fixation. Eye movements provide 'an unobtrusive, sensitive, real-time behavioural index of ongoing visual and cognitive processing' (Henderson and Ferreira 2004: 18) and give us clear and reliable data on the allocation of attention (see also Holsanova, Holmberg and Holmqvist 2008). The basic operation of the eyes in processing information runs as follows.

> When we read, look at a scene, or search for an object,
> we continually make eye movements called *saccades*.
> Between the saccades, our eyes remain relatively still
> during *fixations* for about 200–300 ms. There are
> differences in these two measures as a function of the
> particular task . . . Saccades are rapid movements of the
> eyes with velocities as high as 500° per second. Sensitivity
> to visual input is reduced during eye movements; this
> phenomenon is called *saccadic suppression* (Matin 1974)
> . . . We do not obtain new information during a saccade,
> because the eyes are moving so quickly across the stable
> visual stimulus that only a blur would be perceived (Uttal
> and Smith 1968) . . . As we look straight ahead, the visual
> field can be divided into three regions: *foveal, parafoveal*
> and *peripheral*. Although acuity is very good in the fovea
> (the central 2° of vision), it is not nearly so good in the
> parafovea (which extends out to 5° on either side of
> fixation), and it is even poorer in the periphery (the region
> beyond the parafovea). Hence, we move our eyes so as to
> place the fovea on that part of the stimulus we want to see
> clearly. (Rayner 1998: 373–374; emphasis in original)

So the question becomes: what proportion of time do participants fixate on the carbon footprint information on each of the products, compared to each of the other categories of information represented on the same products? This is on the understanding that this measure of visual attention (with the fovea being directed at each information category) reflects the amount of processing of each of the different categories of information with clear implications for consumer behaviour and the efficacy of the carbon footprint approach for combating climate change.

Of course, the answer to this question won't solely depend on the design or aesthetics of the product. Since the 1940s with the 'New Look' approach to perception, it has been clear that perception is an active and constructive process that often operates in a top-down fashion. Often critical are the needs and values of the perceiver. In a well-known study, Bruner and Goodman (1947) asked children from different socio-economic groups to estimate the size of coins by adjusting the diameter of a beam of light. Children from poorer backgrounds overestimated the size of the coins compared to richer children because, according to the authors of the paper, poorer children place higher value on such coins than richer children and this impacts on their basic processes of perception. This overestimation happens unconsciously and is outside the individuals' awareness or control (see also Balcetis and Dunning 2006; Greenwald 1992). So the fact that many people hold positive implicit attitudes to low carbon footprints may well impact on their patterns of visual attention to the products. As Bowman, Su, Wyble and Barnard (2009) wrote: 'Humans have an impressive capacity to determine what is salient in their environment and direct attention in a timely fashion to such items.' Carbon labelling *should* be salient to participants in this experiment; the question is how impressive is their capacity to direct their visual attention at the carbon footprint information, in the context of competing information about price, nutritional content, value, usage, etc.

Three products were used (see Figures 7.1–7.6): a Tesco low-energy light bulb (part of the 'Greener Living' range), Tesco's 'Non-Bio' liquid detergent and Tesco's own-brand freshly squeezed orange juice. These were photographed on a flat matt background with the bulb's front and back in a single shot as one image (front of bulb to the right, back of bulb to the left – view 1; front to the left, back to the right – view 2, in order to control for natural biases in patterns of left–right looking), the orange juice front and side as one image (front to the right – view 1; front to the left – view 2) and the detergent's front and back as one image (front to the right – view 1; front to the left – view 2).

These images were shown in slide-show format on a Dell desktop computer monitor to ten participants with a 10-second exposure time for each slide (participants were merely told 'to look at the images'). We set up an ASL Model 504 remote eye tracker in the laboratory, in front of the computer monitor on which the stimulus material was to be shown. The eye tracker employs a camera surrounded by infrared emitting diodes to illuminate the eye of the participant looking at a screen. The participant's point of gaze on the screen is determined by the camera combining the position of the pupil and the corneal reflection. The remote camera in the eye tracker fed into a screen for the experimenter's observation of the positioning of camera observing the eye. From a separate computer, the experimenter was able to adjust the illumination of the infrared camera and the 'pan/tilt' of the camera in the eye tracker to enable recognition of the pupil and corneal reflection.

The recordings were analysed using Irfan View and each 40 ms frame was manually coded in terms of participants looking at each of the possible information categories represented on the products or packaging. There were 10 participants × 6 slides × 10 seconds × 25 frames, providing something like 15,000 individual data points which were individually coded and analysed. Examples of the individual frames in the analysis are outlined in Figure 7.7

Visual attention in the first 5 seconds and the second 5 seconds were analysed separately. A third analysis focused on the first fixation of each participant on each slide. A fixation was defined as the eyes remaining still for a minimum of 200 ms – in other words, for five frames or more (see Rayner 1998: 373). In the analysis of the first fixation of each participant, the exact duration of the fixation was also analysed. The scoring of each individual frame with respect to the informational categories was highly reliable.

Now for the details of what we found. In the case of the light bulb, considerable visual attention was directed at the carbon footprint icon in the first 10 seconds (a mean of

82.5 intervals of gaze directed at the icon; in other words participants looked at the carbon footprint icon for a mean of 3.3 out of 10 seconds). This was by far the commonest focus for visual attention in this time period. The second commonest focus was the accompanying carbon footprint information (printed above and below the actual icon), and this competed most closely with other information about the energy backers of the product (EDF energy) and the fact that this product came from the 'Greener Living' range. The information that the bulb lasted for 6 years was also a common focus of visual attention. In other words, in the case of the light bulb much of the visual attention seemed to focus on 'green' aspects of the product (see Figure 7.8).

When these 10-second periods of visual attention were broken down into the first and second 5 seconds, some interesting differences started to emerge. Although the carbon footprint icon was looked at for a mean of 40.0 intervals in the first 5 seconds, and 42.5 in the second 5 seconds (thus showing a remarkable degree of consistency), this was not the case with the accompanying information about carbon footprint. This information was only looked at for 8.7 intervals in the first 5 seconds and 39.5 in the second 5-second period. This showed the most striking divergence in any of the information categories examined. In terms of the first 5 seconds, after the carbon footprint icon, the most frequently looked-at category was the EDF energy label followed by the 'Greener Living' label. It appears, therefore, that in the case of the low-energy light bulb, participants were primed to seek information relative to green issues and they did this more or less within the time frame that they would normally have in supermarkets for viewing products in making their consumer choices (see Figure 7.9).

In the case of the orange juice, the main attentional focus in the first 10 seconds was on the information that the oranges were picked and processed within 24 hours (a mean of 79.8 intervals), followed by the product image (a mean of 64.2 intervals). The third major focus of attention in the case of the orange juice was the price (this ranking excludes the category 'other', which is the residual category). The focus on the carbon footprint icon was lower down the list (a mean

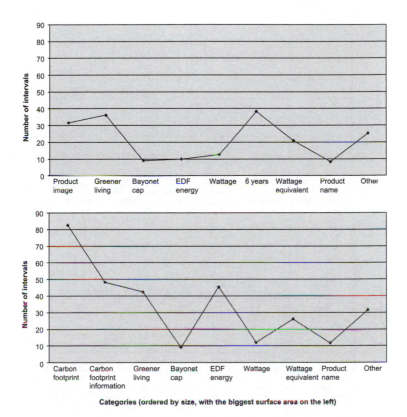

Categories (ordered by size, with the biggest surface area on the left)

Figure 7.8 Number of (40 ms) intervals with gaze directed at each category, for the first 10 seconds of viewing: (top) light bulb, front; (bottom) light bulb, back.

of 33.1 intervals), but was still higher than the attentional focus on the nutritional information (a mean of 27.3 intervals). The accompanying information about carbon footprint was lower down again, competing in terms of attentional focus with things like 'not from concentrate' and the reminder of the price on the side view of the juice packaging (see Figure 7.10).

Looking separately at the attentional focus for the first 5 seconds and the second 5 seconds, again some interesting differences emerged (see Figure 7.11). Unlike the low-energy

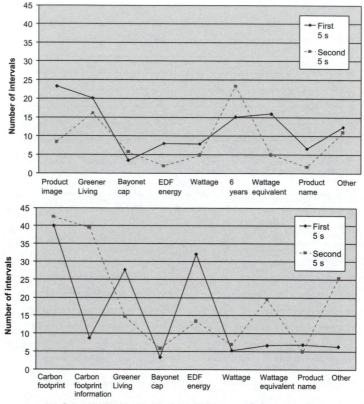

Categories (ordered by size, with the biggest surface area on the left)

Figure 7.9 Number of (40 ms) intervals with gaze directed at each category, for the first and second 5 seconds of viewing: (top) light bulb, front; (bottom) light bulb, back.

light bulb, the carbon footprint icon was typically not fixated for very long in the first 5 seconds, but was only really fixated in the second 5 seconds; indeed, the amount of fixation in the second 5 seconds increased by a factor of 3. However, the accompanying information about carbon footprint did not go up nearly as steeply between the first and the second 5-second intervals. Nutritional information was hardly focused on at all in the first 5 seconds (a mean of 5.8 intervals), representing just over 1/5 of a second of actual

Figure 7.1 Light bulb: view 1.

Figure 7.2 Light bulb: view 2.

Figure 7.3 Orange juice: view 1.

Figure 7.4 Orange juice: view 2.

Figure 7.5 Detergent: view 1.

Figure 7.6 Detergent: view 2.

Figure 7.7 Eye tracking example.

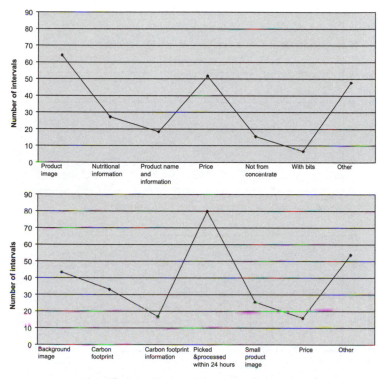

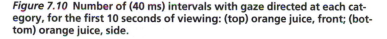

Categories (ordered by size, with the biggest surface area on the left)

Figure 7.10 Number of (40 ms) intervals with gaze directed at each category, for the first 10 seconds of viewing: (top) orange juice, front; (bottom) orange juice, side.

looking, whereas it increased in the second 5 seconds by a factor of 4. So it looks as if in the case of nutritional information people need longer to get round to reading the textual information about nutritional value, but this pattern of increased focus on relevant textual material does not quite generalise to carbon footprint information.

In the case of the detergent, the main focus of visual attention in the first 10 seconds seemed to be that this detergent can be used at 30° (a mean of 80.8 intervals) followed by the product name (a mean of 62.8 intervals) and the product instructions (57.7 intervals). Of course, the

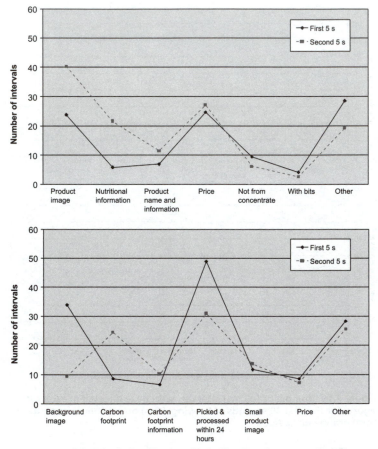

Categories (ordered by size, with the biggest surface area on the left)

Figure 7.11 Number of (40 ms) intervals with gaze directed at each category, for the first and second 5 seconds of viewing: (top) orange juice, front; (bottom) orange juice, side.

temperature at which you wash clothes is relevant to issues concerning sustainability, but it is not the carbon footprint *per se* that is attended to here. The carbon footprint icon (a mean of 4.6 intervals) was looked at less than the fact that the product was suitable for sensitive skin (a mean of 4.8 intervals) and the information that there was 1.5 litres in the container (a mean of 5.4 intervals). The accompanying information about carbon footprint was looked at more, but this did not really compete with the main foci of visual attention for the detergent. Interestingly, in the case of the detergent (but not with the other products), there seemed to be a relationship between the surface area of the representation of the information and the proportion of time spent looking at it, as can be seen in Figure 7.12.

In terms of the possible differences between the first and second 5 seconds of visual regard, again some interesting differences emerged. The fact that this detergent can be used

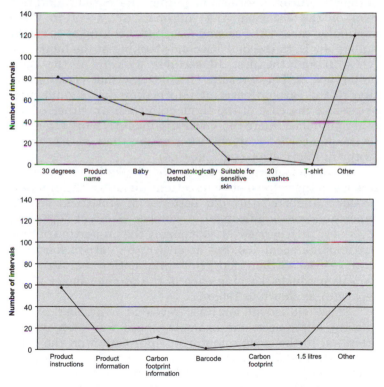

Categories (ordered by size, with the biggest surface area on the left)

Figure 7.12 **Number of (40 ms) intervals with gaze directed at each category, for the first 10 seconds of viewing: (top) detergent, front; (bottom) detergent, back.**

at 30° was looked at in the first 5 seconds, but this dropped dramatically in the second 5 seconds. The carbon footprint icon was looked at more in the first than in the second 5 seconds (and indeed hardly at all in the second period – a mean of 20 ms). Similarly, the carbon footprint information was looked at more in the first than in the second 5 seconds, but not necessarily with the kinds of duration that would entail adequate understanding (see Figure 7.13).

In the case of all three products, statistical analyses revealed that our experimental participants spent significantly longer looking at information not related to the carbon footprint information (either the carbon footprint icon or the accompanying information) except in three particular cases – the second 5 seconds looking at the light bulb (view 1), the second 5 seconds looking at the light bulb (view 2) and the first 10 seconds looking at the light bulb (view 2). These results clearly show that visual attention to the carbon footprint information was significantly greater in the case of the light bulb than on either of the other products, where there was significantly more attention on the non-carbon footprint information than on the carbon footprint information (see Table 7.4).

What has emerged so far is what looks like a clear difference in the pattern of visual attention across the different products (inferred through patterns of statistical significance and non-significance). A number of direct statistical tests were then carried out comparing the proportion of time participants spent looking at either the carbon footprint icon or the carbon footprint information between the various products (keeping each of the product image views separate), as shown in Table 7.5.

These statistical analyses revealed a number of significant differences; for example, there was more visual attention to the carbon footprint information on the light bulb in the first 5 seconds than to the carbon footprint on the orange juice (both view 2). Similarly there was more visual attention to the carbon footprint on the light bulb in the first 10 seconds than to the carbon footprint on the orange juice (again view 2). There was also significantly more visual attention to the carbon footprint of the light bulb than to

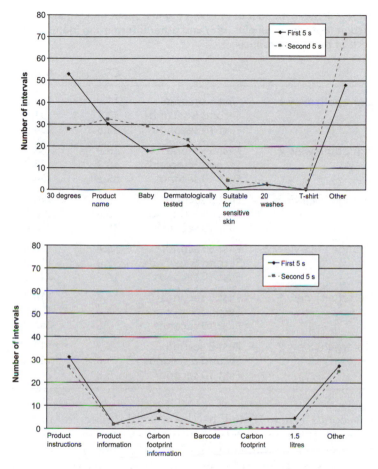

Categories (ordered by size, with the biggest surface area on the left)

Figure 7.13 Number of (40 ms) intervals with gaze directed at each category, for the first and second 5 seconds of viewing: (top) detergent, front; (bottom) detergent, back.

the carbon footprint of the detergent in each of the six individual comparisons carried out.

In terms of the comparison between the orange juice and the detergent there was significantly more visual attention to the carbon footprint of the orange juice in the second

Table 7.4 **Mean number of intervals (40 ms) spent looking at the carbon footprint (icon plus info) and other information (non-CF) for each of the three products**

		CF	Non-CF
Light bulb view 1	First 5 seconds	11.8	113.2
	Second 5 seconds	33.7	91.3
	First 10 seconds	45.5	204.5
Light bulb view 2	First 5 seconds	36.9	88.1
	Second 5 seconds	48.3	76.7
	First 10 seconds	85.2	164.8
Orange juice view 1	First 5 seconds	8.8	116.2
	Second 5 seconds	26.8	98.2
	First 10 seconds	35.6	214.4
Orange juice view 2	First 5 seconds	6.4	118.6
	Second 5 seconds	7.8	117.2
	First 10 seconds	14.2	235.8
Detergent view 1	First 5 seconds	6.6	118.4
	Second 5 seconds	1.0	124.0
	First 10 seconds	7.6	242.4
Detergent view 2	First 5 seconds	5.2	119.8
	Second 5 seconds	3.6	121.4
	First 10 seconds	8.8	241.2

5 seconds (view 1) and in the overall 10-second interval (view 1). However, with the alternative view – view 2 (that is, the front of the product on the left in the photograph) – there were no significant differences. In other words, this array of statistical comparisons revealed a rank ordering in the products in terms of visual attention to the carbon footprint information in the first 10 seconds, with the bulb in first place, the orange juice in second place and the detergent last.

The analysis of the first fixation revealed striking individual differences in terms of where our experimental participants fixated for the first time when they looked at the packaging of certain products. There were also striking individual differences in terms of how long each of these

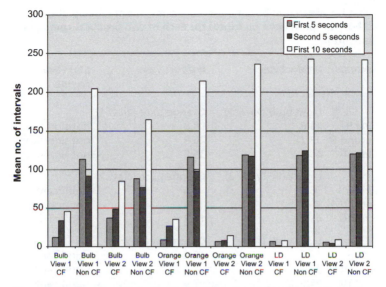

Figure 7.14 **Mean number of intervals (40 ms) spent looking at the carbon footprint (icon plus info) for each of the three products.**

first fixations lasted for. Participant 1 (light bulb, view 1) looked first at the wattage information on the light bulb with a fixation of 200 ms, whereas participant 3 (light bulb, view 1) looked first at the EDF label with a much longer opening fixation of 2.5 seconds. Interestingly, participant 7's first fixation on the light bulb (view 1) was on the carbon footprint icon with an initial fixation of 480 ms. With 10 participants and 6 slides there were 60 initial fixations to consider and out of those, only 4 were on the carbon footprint icon (and none were on the accompanying carbon footprint information). In other words, in less than 7% of all cases did participants fixate immediately on either the carbon footprint icon or the accompanying carbon footprint information when they looked at the packaging of products in which the carbon footprint was clearly labelled.

So what conclusions can we draw from all of this? There is a major argument proposed by many prominent individuals that one way of tackling climate change, and halting the year-on-year increase in greenhouse gas emissions, is to

Table 7.5 **Mean number of intervals (40 ms) spent looking at the carbon footprint (icon plus info) for each of two products compared statistically**

Seconds	Product view	Product view	Statistical comparisons
	Light bulb (view 1)	Orange juice (view 1)	
First 5	11.8	8.8	not significant
Second 5	33.7	26.8	not significant
First 10	45.5	35.6	not significant
	Light bulb (view 2)	Orange juice (view 2)	
First 5	36.9	6.4	significant
Second 5	48.3	7.8	not significant
First 10	85.2	14.2	significant
	Light bulb (view 1)	Detergent (view 1)	
First 5	11.8	6.6	significant
Second 5	33.7	1.0	significant
First 10	45.5	7.6	significant
	Light bulb (view2)	Detergent (view 2)	
First 5	36.9	5.2	significant
Second 5	48.3	3.6	significant
First 10	85.2	8.8	significant
	Orange juice (view 1)	Detergent (view 1)	
First 5	8.8	6.6	not significant
Second 5	26.8	1.0	significant
First 10	35.6	7.6	significant
	Orange juice (view 2)	Detergent (view 2)	
First 5	6.4	5.2	not significant
Second 5	7.8	3.6	not significant
First 10	14.2	8.8	not significant

empower consumers to make informed adjustments to their patterns of consumption, by providing them with relevant and accurate information about carbon footprint. The fact that many people do seem to have a strong positive implicit attitude to low carbon footprint products lends some credence to this general view. People would seem to be (already) primed to change their behaviour. A number of retailers (including Tesco in the UK) are now selling products with carbon footprint information clearly marked on their own-brand products to allow this empowering process to commence.

But the interesting psychological question is to what extent the carbon footprint information (usually consisting of an icon plus accompanying textual material) successfully directs the consumers' visual attention to itself, in competition with all the other information that appears on the packaging or on the products themselves. Detailed analyses of the recording of each participant's pattern of looking (each 40 ms frame was manually coded) during two 5-second periods of regard revealed that with certain products significant amounts of attention were directed at the carbon footprint and it did occur within the first few seconds. In our research, which compared three different products, most visual attention was directed at the carbon footprint of a low-energy light bulb compared with the carbon footprint of a carton of orange juice or a container of detergent. Least visual attention was directed at the carbon footprint of the detergent. In the case of the light bulb, attention was directed within the first 5 seconds at the carbon footprint icon, but attention only moved to the accompanying textual material in the second 5-second period (with only minimal attention in the first 5 to this textual material). It seemed to take much longer for participants to attend to the basic carbon footprint icon in the case of the orange juice (only really appearing in the second 5-second interval), and in the case of this product they hardly attended to the accompanying information at all. In the case of the detergent there was minimal visual attention to any aspect of the carbon footprint.

Bowman, Su, Wyble and Barnard (2009) recently wrote that 'Humans have an impressive capacity to determine what is salient in their environment and direct attention in a timely fashion to such items.' Carbon labelling on products, which some see as a major part of the solution to the issue of climate change, should surely be a salient part of all of our everyday lives, but it seems that it is only the carbon footprint of *certain* products that is really salient (at least within the critical 5-second time frame of everyday supermarket shopping). Our research also found that the carbon label (the carbon footprint icon plus the accompanying information) was the focus of the *first* fixation of our participants in only about 7% of all cases (and only 10% of the first fixations even in the case of the low-energy light bulb). In other words, the carbon label is not where participants look first. From a psychological point of view the only way that carbon labelling will ultimately work is if the information is designed in such a way that it does become the primary focus of visual attention in the first few seconds. At the moment this appears to occur only with certain products, which we already associate with being 'green'. How we redesign products to make this information stand out more thus becomes a critical issue in the fight against climate change, as does the issue of how we go about making carbon footprint information more emotionally salient to people generally. The reason for this is that we know that emotional valence, and our values more generally, affect our perceptions of the world (Bruner and Goodman 1947). They even affect the moment-to-moment unconscious eye movements that are the crucial building blocks of this process of perception (and also, of course, the subject of our own research).

The overall implication is that if we are to combat climate change by providing consumers with carbon footprint information, then we will need to consider much more carefully how to make this carbon footprint information significantly more salient, because if the carbon label is not 'seen' in the right time frame, then it simply cannot be effective. This process will involve not just changing the packaging of products (essentially a design issue, but guided by psychologists who understand the limits of time-dependent cognitive processing) but also (and somewhat more dauntingly) it will involve changing certain aspects of the fundamental psychology of consumers (and not just their implicit attitudes, which already seem positive), because salience (as we all know) really is in the eyes of the beholder. So until we change crucial aspects of the psychology of the beholder, none of this activity on carbon labelling will really work, despite the very best of intentions.

PART II

Notes on habits

8

Eden reclaimed

I was slowly walking back to my hotel room, stepping carefully around snails with large, almost comical heads, and brightly coloured shells. They were the size of small rotund sparrows. It was 1.30 a.m., the end of another day in paradise. I was staying in a beautiful hotel with manicured verdant lawns that swept down to the white sand and clear warm waters of the Indian Ocean. It was now January and I was attending a sustainability conference in Mauritius to present my new research (of course, I saw and *felt* the irony of this).

I had spent the day in various plenary sessions listening to the latest views on environmental sustainability. What was nice about the conference was that it had a type of session called 'garden conversations' which were unstructured 60-minute sessions that allowed delegates to meet the plenary speakers and talk with them informally about any emerging issues and, in addition, there were 'talking circles' which, according to the organisers, were 'meetings of minds, often around points of difficulty. They are common in indigenous cultures. The inherent tension of these meetings is balanced by protocols of listening and respect for varied viewpoints. From this, rather than criticism and confrontation, productive possibilities may emerge.' I had presented my work on the implicit attitudes to carbon footprints and the response was very favourable, many productive possibilities emerged, and I discovered that few researchers interested in sustainability seemed to have thought about this issue before but I could see that many were thinking about it now.

I had chosen a hotel close enough to where the conference was being held just outside Port Louis, in Pointe aux
Sables, but really miles apart from the chaos and patchy
squalor of that town. I was guided to my room by the sound
of the crapaud, the Creole name for the little frogs crying
out for a mate in the ponds dotted around the hotel.
These were man-made with slate waterfalls, with the water
emerging out of the mouths of golden lions, and goldfish
swimming under the gentle sparkling falls, but the tropical
nature of the island had invaded the interconnected ponds
to give them new life; the whole thing pulsated like a
membrane. The croaks sounded like the raspy death rattle
of a human being, that terrible Cheyne-Stoking sound and
rhythm, but with the opposite emotional significance. This
noise was all about life and the celebration of libido rather
than the celebration that thanatos had finally neared the
end of its fateful journey. I timed the interval between the
croaks – almost exactly one per second. The whole hotel was
pulsating with this incessant noise that seemed to be getting
louder. The volume bore no relation to the tiny creatures
from which the noise comes. I picked one up to examine it,
and it sat quietly in my hand before I released back into the
lush pond life.

I had travelled halfway across the world to be here. One
and a half hours to Charles de Gaulle airport and then
eleven and a half hours to Sir Seewoosagur Ramgoolam airport. Then, I had to travel another one and a half hours
across the island in a private taxi, sitting there with its
engine running so that the air conditioning would make it
nice and cool for some Westerner like me. Of course, I could
see the extraordinary irony in holding a conference on
sustainability in Mauritius with academics reluctantly
travelling from Minnesota and Harvard and Manchester
to be there. But then again it does not pay to be too ethno-
centric when it comes to academic debate. Sustainability is
a global issue, and Mauritius is much closer to Africa and
India and Australia than many other possible locations. And
Mauritius did give me a sharp reminder, with its corrugated
iron huts with hardboard walls, and buckets on the ends
of rope for toilets, of the sheer economic significance of

premier tourism in countries like this, with hard-working Western executives with lots of air miles in need of air-conditioned taxis and hotel rooms. If, and when, we cut back on air travel to Mauritius and other jewels in the Indian Ocean, either voluntarily or through some kind of pro-hibitive legislation in the future, these places will suffer greatly. And this is a real possibility in terms of the current ethos, with air travel being often singled out for particular criticism in terms of global warming. Walker and King (2008) say that 'In terms of overall numbers that's not strictly fair.' But they add:

Aviation is directly responsible for about 700 million tonnes of carbon dioxide each year, which is just 1.6 per cent of global greenhouse emissions. However, molecule for molecule the emissions count for much more than they would on the ground because planes are very efficient at causing greenhouse warming. High-altitude deliveries of nitrogen oxides (which form ozone, another greenhouse gas), as well as the water in contrails that can go on to form cirrus clouds, together enhance the direct effect of carbon dioxide by up to a factor of three. (2008: 124)

Sustainability is all about choice, but many of the choices are not easy. The next afternoon I lay in a hammock overlooking the sea, watching a weaver bird pick at the bread that I had dropped: the bread dropped carelessly rather than intentionally. Two small familiar-looking sparrows approached to nibble the bread beside the weaver bird; a red-crested, red-bodied serin walked arrogantly towards the other birds and stood in the middle waiting his turn. A water vole with pink feet ran through the scene and appeared to squeak. Two tiny grey doves arrived and walked around each other in what looked like a dance. This small part of earth, covered in luxurious thick grass, was fully alive and per-fectly harmonious. The sea lapped the shore and I started to drift off; this was as close to a tropical paradise as I have seen or I could imagine. In the blue distance two fishermen a half-mile out from the shore cast their lines. They looked as if they were walking on water. But the perfection of the whole scene made me worry in the way that things do when they are too good.

I reminded myself what Erich Fromm wrote in 1941 about the first choices facing man as described in the Book of Genesis. Fromm was here attempting to understand the destructive political forces that were then gripping Europe in the form of the rise of fascism. 'Man and woman live in the Garden of Eden in complete harmony with each other and with nature. There is peace and no necessity to work; there is no choice, no freedom, no thinking either. Man is forbidden to eat from the tree of knowledge of good and evil. He acts against God's command, he breaks through the state of harmony with nature of which he is a part without transcending it' (1941: 27). This, according to Fromm, was the first act of human freedom, the first transition from mere unconscious existence, the first essentially human act. Human beings are just that because they do have choices, and the consequences of these choices are the parameters that define all our lives.

I lay there pondering some of our choices for the future. So what exactly were they? Was it just going to be air miles for the elite, the thinkers like the ones at this conference, conspicuous consumption for the few, and a new silent hypocrisy (where we simply won't mention any longer what we consume – but we will consume it nonetheless), the rebirth of a new elitism that will even be acceptable to someone like myself? Were they the kinds of decisions that we will have to make for the good of the planet because we all can't go jetting off across the world to hold a dialogue in a garden circle on a tropical paradise?

I went for a run that afternoon and soon saw, more starkly this time, that this paradise that I was now inhabiting was something of an illusion, a mirage created by unconscious forces, driven by images of white sand and palm trees and glamour in a desert of crippling poverty, with descendants of generations of African and Indian slaves that worked on the sugar-cane plantations that now provided the impeccable service culture, with dark-skinned waiters and waitresses in sailor suits and Nehru jackets that appeared to be naturally obsequious (but that was a dangerous conclusion although their smiles did fade naturally and slowly as they turned away from you as if they were genuinely pleased to serve you). I was suddenly overcome by an ill-defined and unfocused guilt. I felt guilt about the beauty of the world, or that part of the world that I was now privileged to inhabit, and guilt about my position in it, in that I still did not appear to be doing much to help even some bits of the planet (as if you could choose to help just bits of it rather than the whole natural thing). I did, however, make a list of things that I could do while I was here to help assuage this guilt. I came up with a list of three.

1   I will in Mauritius only travel under my own steam – except at night or when I have got heavy bags – and I will not eat more to fuel it. This despite the fact that my own steam really means running and my runs are taking me past the dry, flattened, frog-like skin of many dead rats who have feasted on the sugar-cane and have then been run over by fuel-inefficient cars as they venture out of the fields of sugar-cane in their bloated state. I also noticed on these runs that nobody was walking outside the towns, cars were everywhere, and plastic bags of rubbish were dumped beside the roads. There is not much of a green conscience when you are this poor. The cars and lorries came far too close to me and I was continually stepping into spikes of sharp cane to avoid being knocked over.

2   I must reduce the carbon footprint of what I eat. This is, I now know, in line with my general unconscious instinct, even if it is not as strong as it might be, but for some reason it is still not being translated into practice. I will eat only local produce at the hotel and I will quiz the staff diligently to determine whether the pineapple or the pawpaw is more local to the north west coast of Mauritius.

3   I will not leave the TV on standby, even though the red button acts as a constant reminder to switch on the BBC World News to keep up with Israel's invasion of Gaza because I need to always remember how politics can change everything in an instant, unleashing

darker unconscious forces barely restrained in a
civilised world (perhaps rereading Fromm had made me
more sensitive to detecting the work of the destructive
death instinct, which was now seemingly being allowed
free rein in the Middle East).

I laughed at the first two on my list and felt nothing but
despair with the third. Every Israeli spokesperson seemed
to be a softly spoken woman, in uniform or not. Western
reporters were being prevented from entering Gaza, and they
reported instead from the Israeli side of the border and
therefore often had direct footage of the effects of the small
number of rockets still fired by Hamas into the border towns
of southern Israel. The description of the damage these
rockets were doing always seemed to precede the horrendous
destruction of Gaza.

I switched channels and watched Reading play Watford
in the Coca-Cola Championship. This match reminded me
that this is a very small world with shared concerns, and that
maybe some degree of convergence is possible. The crapauds,
however, getting faster in frequency and even higher in
volume, and now sounding like a thousand football rattles,
suggested otherwise. I retired in a black mood, thinking of
how strange the crapauds sound, and that the world is an
odd, diverse and varied place in which everything changes as
you move around it, there is no fixed reference point anywhere and consensus on anything – war, famine (and who is
to blame), sustainability, global warming – will be virtually
impossible.

The next morning the sun was shining again; it was high
in the sky, brilliant and intense, and my spirits lifted
immediately. In front of my veranda was a panoply of palm
trees with their heavy yellow coconut burden. Incongruously
a rose bush sat to my left bearing pink roses, presumably for
the honeymoon couples that come here. The palm trees were
thick and luxuriant and spaced at a pleasant distance but
with no obvious pattern, as if nature has been tamed and
improved by man's intervention. They had even made the
spacing seem random and natural, but it was far too pleasing
to the eye for that.

But at midday suddenly everything started to change again. The rain was coming once more, but now it felt different. However, this was also the day of my first sailing lesson and I did not want it to be spoilt by the wind and the rain: after all, sailing needs wind doesn't it? The attendant in the sailing office advised me to leave my powder blue Armani sunglasses in the kiosk – 'In case you capsize or are knocked unconscious by the boom, monsieur,' he said helpfully in his French creole. I also offered up my grey Nike Pegasus running shoes stained with clay and reeking of sweat and effort. He motioned to the step below the kiosk: 'Leave them there monsieur, they'll be quite safe.' He obviously didn't want to touch them. The instructor stood forlornly on the small laser sail boat. 'The rain,' he said, 'it is coming. It is a cyclone.' I put on my yellow lifejacket carefully, fastening each of the three poppers in turn; I didn't want to miss this opportunity.

'Cyclone is good,' I said, 'lots of wind.'

'No, no, monsieur, too much wind, cyclone blow, blow, blow.'

He obviously wanted to go back into the kiosk to play cards with his friends. His facial expression was telling me everything that I needed to know.

'Do you suffer from panic?' he asked.

'Pardon?' I said.

'Do you have panic? Because if you do have panic we can't do sail, if you panic we drown. We might as well go home.'

I shook my head slowly and mournfully, eyes wide open like a small but determined puppy. 'I don't do panic,' I said. We walked slowly together to the slightly built craft and I clambered in clumsily, which only seemed to worry him more.

'Cyclone is very bad,' he added, allowing me one last chance to back out.

We sat on the small boat facing each other as he explained what each of the small instruments did. The rain was definitely getting harder. There was a thin nylon rope, blue and white, for tightening the sail, a light aluminium rudder for steering the boat and a boom that shuddered unpredictably and violently across the boat.

'You must sit hunched up and duck when the boom comes, otherwise you will be knocked out,' the instructor explained.

We set off into the white silver bay, now grey from the rain. I was gently guiding the boat. He was standing on the bow holding onto the sail, the pelting rain streaming down his face; he was looking miserable, he would clearly rather not be here at this moment in time but I had been insistent. He looked past me towards the shore, wistfully.

He tried one more time. 'The rain,' he said, 'it is coming. It is *très mal.*'

We were picking up speed and the small boat was gliding through the waves. I wanted to take all of the stunning scene in, but the driving rain was forcing my eyes into a narrow slit; I was just focusing ahead.

'Where are you from?' he asked. 'Manchester,' I replied. Soon we were talking about Ronaldo and Alex Ferguson and the way that Manchester United always seem to start slowly in the Premiership but nevertheless always seem to end up on top. We were making contact at last. He asked what I did for a living and when I told him I was a Professor of Psychology he became animated.

'My mother do psychology too,' he said. 'She reads the tarot cards and she reads the rice grains. I have seen her with my own eyes cure a man in a wheelchair using the tarot cards. She found out what the matter with him was and then fixed it by sacrificing a hen. Can you cure the man in the wheelchair?'

I shook my head sorrowfully. 'Unfortunately not,' I said. 'I once cured a student with a fear of public speaking by forcing her to talk to a lecture theatre full of two hundred students, but that was about it.'

'You very cruel man,' he said. 'My mother can walk on coal and she can do sacrifices, just small sacrifices, chicks, hens, goats, nothing too big. She just cuts their head off and reads the tarot. Can you walk on coal, monsieur?'

'No,' I replied, 'but I did walk across the beach this morning and it was very hot indeed!' He didn't really get the joke but I laughed anyway, which he seemed to find uncomfortable. He had a professor with him in his small boat in a developing cyclone who appeared to be laughing at himself.

'Turn left monsieur, left, you must dance with the wind, not fight it. You must learn to live in nature, not fight against it all the time.'

The wind was now much stronger and the boom was being shunted abruptly from side to side; I could see no pattern in the changing turbulence. He, however, was standing on the bow reading the direction of the waves. He clung onto the sail, delicately balanced in the driving wind and rain. Suddenly he shouted out a warning but it was too late. 'Duck, monsieur, duck!' And the inevitable happened: the boom struck me with some force in the back of the head giving me an instant headache but simultaneously waking me up. I suddenly had the first clear thought of the day. What was I doing in the middle of the Indian Ocean on a tiny boat (almost like a lollipop stick) with the son of the local witch doctor? The rain was torrential.

'What you think of black magic?' he suddenly asked.

'It depends on how it is used,' I replied.

'Sorry,' he said back, sounding as if he assumed that I hadn't heard his first question. 'What you think of black magic?' he asked again.

'Good,' I replied this time.

'You like?' he said, smiling with large white teeth.

'I really like,' I said. 'I would love to meet your mother. Could we perhaps go back to the shore now?'

'No, not yet,' he said, his smile temporarily leaving him, 'your lesson is not up yet. I want to learn more about your psychology. Turn the other way.'

'Towards the cyclone?' I asked hesitantly.

'Yes, that way,' he said, gesturing towards the cyclone but without actually mentioning it.

I sat in silence for a moment just listening to the rain bouncing off the boat.

'Have you met Ronaldo?' he asked, and looked disappointed when I failed to answer. 'Would he like a sacrifice? What about Alex Ferguson? My mother would sacrifice a large goat for him for Manchester United to beat Chelsea. You tell me what score you want, two nil, three nil, my mother arrange it.' His mind was racing away with him, thinking about all of the commercial possibilities when east meets west. I just wanted him to keep focused and get us back safely. I was trying to put all that I had learned in this lesson, which was not very much, into practice and the small white sail-boat responded well, sprinting back towards the land like a dog trying to get out of a storm. I shook hands with him when I half fell out of the boat onto the shore. 'Thank you,' I said.

The rain was torrential now: torrents fell through the trees, torrents raged though the rivulets and drains. There was a large cyclone off Rodrigues Island maybe heading south east, maybe turning towards Mauritius, but it had still sent all this rain here. I stood sheltering with the beach boys who hired out the pedalos and the sail-boats to the tourists. They told me that the cyclones were far more frequent and more intense than they had been in previous years, and that the beautiful manicured grounds of this hotel that sweep down to the lapping waves of the Indian Ocean were now often the victims of flash floods. They also told me that the temperature was rising – they had seen this with their own eyes. Even though they were all in their late teens and early twenties they said that they personally had seen the change – each summer now the temperature climbed to 37 degrees rather than peaking at 32 degrees as it did when they were younger. 'It's getting too hot for some tourists,' said one dark African boy with a fashionable goatee. 'And much, much too wet,' said another. 'One day the tourists may not want to come.' I knew that they were right from some recent studies I had read. But it is one thing reading it in a paper, and quite a different thing to stand there feeling the new intensity of the cyclone on the horizon.

I climbed back off the beach and saw that the palm tree outside my room now appeared to be emerging out of a deep pool about two metres across. It was an odd sight, like an island that had been drowned – maybe like a vision of the future. A little bird, flame red in colour, called a red cardinal, hopped towards the pool of water around the base of the tree to pick insects from the wet soft ground around the new pond. The heavy patter of rain through the leaves was everywhere. It was such an evocative image, an image that I knew would prove to be indelible, and it pushed one clear thought my way. It was such a cruel thought. The thought was that we helped make this Eden, we tamed it, we structured it and we imbued it with our unconscious desires and our symbolism of tropical island beauty and romance and sanctuary and aspiration, but then our choices – one at a time, linear, sequential and unconstrained – might already have helped to bury it.

9

Old habits

Sometimes you sense when it's time to change. But the instigation of new actions can be a difficult and, on occasion, uncomfortable process for any individual. It may require deliberation when previously there was none, and the interruption of automatic unconscious routines to be replaced by something more conscious and controlled. Psychologists and other social scientists often talk about 'habits' and argue that we will save the planet only if we do something about the destructive and selfish habits of all of us. But I always find the concept of the 'habit' mildly disconcerting. Of course it reminds us that habits are forms of learned behaviour, not instinctual and not *necessarily* biologically programmed (although some may well be). The discourse about habits talks about 'bad habits', 'breaking a habit' and acquiring 'new habits', which is both positive and empowering. But my issue with habits is that I just do not see them as more or less behavioural accidents, reinforced in childhood by a contiguous maternal smile or by the sheer contingency of a sibling's quiet look. They are more deeply ingrained than that, critically attached to aspects of the individual's personality for sometimes inexplicable reasons. They are often bound up with something much deeper than chance behavioural contingencies, the way that some behavioural psychologists might have us believe.

The image of the palm tree outside my room in Mauritius, its roots fighting for survival in its new private lagoon, made me want to think about some of my core habits – the habits that define me but may be destructive forces from the point of view of the planet. One of these core habits, now that I am in confessional mode, is my sheer level of consumption and indeed my whole relationship with possessions. I buy far too many clothes, which I accumulate and hoard. These clothes and sports equipment and running shoes are part of me, and I hang on to them for years until every chest of drawers and wardrobe is full to bursting with the internal rails bowed and on the verge of breaking from the sheer weight of the clothes being hung on them. There is no room for anything else and yet I constantly buy. 'I will find a space,' I say to another set of disapproving eyes. Suits jut outwards from the wardrobe suspended from the handle in unsteady configurations; some shirts or suits disappear in the wardrobe and I will find them a year or two later, brand new but crumpled and creased and almost unwearable.

A few years ago I treated a shopaholic for a television series for the BBC: her name was Carmel and she lived in a bungalow just outside Derry in Northern Ireland. She was like me in many respects, at least in terms of her shopping habits. She had hundreds of items of clothing and maybe a hundred pairs of shoes. The clothes were stuffed in drawers, under all the beds in her house, including her parents', in the wardrobe in her boyfriend's house and even in the back of her car. It was her parents who rang the programme and as part of the 'therapy' I made her retrieve the clothes, often still in their wrappers, from the wardrobe and the drawers and under the beds and place them in the front room of her bungalow, eventually forming this giant haystack of clothes, then I sat on top of this clothes mountain with her and I interviewed her on camera. She was a lovely girl and very relaxed but it still felt uncomfortable, because I could have been interviewing myself. I told her off camera that I was no different from her and that I had the same fragile ego that needs that approving look which can most easily be elicited by a new set of clothes. (If you try to get that same approving look in old clothes, or even just clothes that you have worn before, you are more likely to get rejected for being too insecure and needy. With new clothes, however, you can turn on the tap of narcissistic supply and let it pump its soft, velvety liquid all over you.)

As part of the programme we fitted a heart-rate monitor to Carmel to see where the excitement of being a self-confessed shopaholic came from. Carmel could have got the buzz from the power of the credit card, or with the inter-action with the sales staff or even by walking down the streets of her native Derry laden with designer bags, all attracting envious glances. But no, the buzz came, and her heart rate peaked, when she got home and tried on her new clothes in front of her family and particularly in front of her beautiful sister, because for a moment all eyes were on Car-mel. It was easy for me to predict what the critical moment would be and to understand exactly what her consumption did for her, her ego and her life. It sometimes pays to be a flawed psychologist.

But from the point of view of the planet I cannot go on consuming in this way because the labels on all of these shirts and suits that I buy tell me that many of them are manufactured in China, competing to be the world's largest emitter of CO_2 and the new bogeyman of climate change. As Walker and King (2008) put it:

> The pace of development in China is extraordinary. A year or so ago, China was building a new coal-fired station a week. Now it is more like two a week and counting . . . their new power stations are especially bad news from a climate perspective because coal is the dirtiest of all fossil fuels, producing not just smoke and smog in the cities, but also much more carbon dioxide for every unit of energy than either oil or gas. (2008: 199)

But of course, life is never that straightforward: you do have sympathy with a government trying to do something about the economic gulf that exists in that country, with vast and conspicuous wealth centred in Beijing and Shanghai and yet 700 million or so people living on less than two dollars per day (see Walker and King 2008: 200). As Walker and King (2008: 200) point out, 'China can say, with justice, that unlike the industrialised West, it has done almost nothing to create the climate problem, and that its citizens play on average a very meagre part in perpetuating it.' China might have a total annual emission rate of 6,467 megatonnes of CO_2, but its per capita emission rate is 5.0 tonnes per person, compared with 7,065 megatonnes for the USA but 24.0 tonnes per person (with the UK at 656 megatonnes and 11.0 tonnes per person).

I need to pass on some of the things I have bought to cut down overall consumption to slow down the pollution from China and other developing countries, but I also need to break the emotional bond between me and my possessions and to separate who I am from what I own, but I knew that this was going to be easier said than done, even when life was conspiring to help me.

In Mauritius the Hotel Maritim had a gym, and every day without fail at roughly the same time I would be there. This is also a big part of being a narcissist. The gym was quite quiet and most days it was just the instructor and me – he was a broad-shouldered Mauritian of Indian descent with a shaved head and a left eye that seemed to be permanently bloodshot, perhaps as a result of the effort he was putting into his bench presses. We would train in parallel and thereby developed a sort of bond, an intimacy that comes from routine and dedicated activity. Each day he would ask me about my runs and often he would remark on the quality of my running shoes. One day I happened to comment that in England I had maybe sixty pairs of running shoes. And from then on he kept asking me when I was leaving and whether I would be leaving in the morning or the afternoon. It was as if he didn't want to miss my departure, although I couldn't really understand why. But then he came right out with it and he asked me whether I would give him my running shoes when I left. 'They are of much better quality than the ones we get here in Mauritius,' he said. I glanced down at the shoes he was wearing: they were also Nike and I could see that they were much bigger than mine. I pointed this out to him and he explained that the shoes he was wearing were several sizes too big for him but that they had been given to him by a previous guest.

So this was my essential dilemma. I know I need to break my habits of consumption and to do something about my emotional attachment to possessions. I know I need to recycle not just tins and cans but my shirts and suits so that hundreds of other shirts and suits don't have to be produced in the first place. Giving a pair of trainers away would be a start, a small moral act that would make me feel better about myself and might be the first step in my attempt to break one of my destructive habits, but it was never going to be easy. I sat that night full of self-reproach, raging at my lack of will. My problem is that I know that this is one of my stable and enduring traits – rooted in the insecurity of my working-class childhood and most certainly not reinforced and conditioned by those around me. This attitude to my possessions has all the wrong cogitations for a mere 'habit': the wrong aetiology, the wrong sustaining features and the wrong connection to my essential self.

My attitude to my possessions has always been there as far back as I can remember; indeed, one of my earliest 'flash-bulb' memories of my childhood centres around the destructive aspects of this attitude. Like all flashbulb memories this is something that I cannot forget, and I do try to forget it because it is an image and a narrative associated with shame, but my conscious will to forget cannot undo what has been stored unconsciously and involuntarily and, it would seem, for a lifetime. The memory concerns a visit to my uncle and aunt's house. They lived in Lesley Street in Ligoniel, at the edge of Belfast, and some Saturdays I would take my box of cornflakes folded over at the top, and my pyjamas, and go up there to sleep between my aunt and my uncle in a house that smelt different from ours. In the morning they would sometimes let me go and play on the steep hill at the end of the street, the steep hill where my fort ended up.

My father made a fort for me at work. He was a motor mechanic for Belfast City Corporation and worked on its buses in the Falls Road depot. He told me that he was bringing something for me and I waited for him for an hour at the bus stop at the top of Legmore Street. It wasn't my birthday or anything like that; it was just a present. I saw him in his oil-stained overalls with his glasses on, getting off the bus with something large wrapped in newspaper. He was a slight man and he could hardly carry it, but he was smiling, because he knew that I would be very pleased when I saw it.

He tottered as he held it in front of him. He wouldn't open the present until we got into the front room. Our dog Spot was jumping all over the furniture, sniffing the paper and barking, shredding the paper with his sharp teeth, too excited. The package was soon opened by the dog and me. I stood staring at the present. I had never seen anything like the fort before. It had brown metal ramparts with zigzag steps shaped out of a single piece of aluminium and a hardboard base. The whole thing was solid and well put together. It must have taken months to make in his spare time at work, and every bit of metal had been shaped by hand. I had received an expensive Christmas present that year – a rocket and missile base, in which the rocket and the missiles both fired. But the fort was different. In these rows of identical mill houses all crouching in that hollow below the hills that ring Belfast, street after street of them as far as the eye could see, all with their cheap identical flowery settees bought on credit from the same shops at the bottom of the Shankill, and the same pictures on the wall of foxes, fawns, infants – in fact anything with big eyes professing innocence and adoration – there was something individual and unique about the fort, made in and for love. And made for me.

I had hundreds of soldiers in a large rusty circular tin that stayed in the damp back room – cowboys and Indians, Confederates and Yankees, knights whose legs and arms moved and could be swapped over, called 'Swap-its', Russian soldiers with a red star in the middle of their grey winter hats. I had bought the six Russian soldiers in Millisle and played outside Minnie McFall's caravan in the sand. The knights on horseback were so intricate and such a delight to look at that my mother put them on display with all the best china in the china cabinet. Two knights, the Red Rose of the House of Lancaster and the White Rose of the House of York, were on parade on the top shelf. We weren't allowed to go near the china cabinet, or feed the gas meter which was just behind it. And when you wanted to play with the knights you had to ask for the key and remove them with a very steady hand from the glass cabinet, which always seemed to tremble and shake with all that china. I can still remember the smell of the china cabinet. All the smells that I have experienced in this life, the lavender in the quiet fields beyond Sainte Maxime, the close-up smell of drying sea-weed on those wild gull-squawking shores north of Santa Barbara, the fragrant smell of leather in the market in Hammamet with mint tea in the background, but I can still smell the inside of that cabinet with greater ease and with greater clarity than any of them, as if I have just leaned into the cabinet and breathed again. Don't ask me to describe the smell; it must have been some kind of cleaning material that had evaporated in a glass container over many years.

All the soldiers eventually found their way into the fort. It was a generic sort of fort though it looked like Fort Laramie, from the Wild West, on the black-and-white television. But my mother told me that her daddy had been to a fort like that in India. George Willoughby she called her daddy, just in case I thought it was George Bell: my cousins Myrna and Jacqueline's brother. My father knew of George's days in India and this might have given him the idea that guided his craftsmanship. British soldiers of the Raj patrolled that fort at night in the corner of our front room, but no shops seemed to sell models of the enemy, whoever the enemy were. But that didn't matter to me, a boy with a fertile and vivid imagination, in a damp, crumbling mill house, who could spend hours on the floor and not be bored. I often didn't want to go out. I was happy in there: sometimes my mother had to make me go out and play on the street.

But one Thursday in July when I was about eleven we were going to my Uncle Terence's and my mother told me that I was now too old for the fort and the soldiers. She was tired of cleaning my kneecaps with Vim because of the amount of time I spent on the floor. 'You're too old to be on your hands and knees all the time. You're too old for that sort of childish nonsense.' It was all done in a matter-of-fact sort of way, as if it was no big deal. The fort got in the way in such a small house. We kept it in the back room, where the wallpaper hung in great damp swathes from the slimy green wall with the damp running down in rivulets. The fort was going rusty like the metal container with the soldiers, like the tools we kept there, like everything else in the house. It had to go and it was loaded into our car. I don't know who loaded it into the car, perhaps my brother. I was told that the poor children up in Ligoniel would love it. I was told that I had had my enjoyment. It was somebody else's turn. I was assured that the children up in Ligoniel weren't as well off as we were. They had no missile sites, or garages with lifts that could be wound up, or forts made at work by their fathers in good jobs. I knew that they were from big families, families sometimes with no work, Roman Catholic families. 'Too bloody idle,' our neighbours liked to say when Big Terry wasn't about (I learned later that my Uncle Terence was himself a Catholic, and that, of course, was a big issue in those days in Belfast, although some days looking back now I can't really understand why).

There was a steep hill at the end of Lesley Street; we called it 'the dump'. I suspect that it wasn't an official refuse site. I remember old settees with rusty springs sticking out and bags of open tin cans with large black crows picking at them. My mother told me to leave the fort out on the dump. She told me that it would be found, and that one of the boys from Ligoniel would have a childhood filled with imagination because of that fort: the fort that my grandfather had fought in, and my father had made.

My uncle came with me as I laid the fort out in the middle of a hill of refuse. Dust and hairs, human and dog, filled the cracks, ingrained and dense like thread. But it was well looked after. That's another expression my family liked. The fort, the car and the front step that my mother would wash every couple of days on her hands and her knees, a white froth on the pavement outside the house swept away by basins of cold water. All well looked after, all cared for. Loved, if you like. It was a very Ulster Protestant way of thinking about these things.

So I carried the fort and left it in the middle of this long slope filled with human debris. A beautiful handcrafted artefact that had been at the centre of my childhood: that still was at the centre of my childhood. That perhaps was the problem. My mother decided that at eleven I shouldn't be in the front room on all fours with cowboys and Indians and Russians with the red star on their caps. I walked back up the slope with my uncle, who talked about our dog being humiliated by a rat in Barginnis Street. 'That dog of yours can't bloody well fight,' my uncle said. 'It's embarrassing. It's not a dog at all.'

We all sat in my uncle's front room, my Aunt Agnes, my father, my mother and Terence's mother. There was a crucifix on the wall as you came in. I had only ever seen one in the Rocks' house. I never understood what it was doing in my Uncle Terence's house before Kevin Rock explained. My mother always said that it was something to do with Terence's mother. I didn't know what though. But I couldn't stop thinking of the fort. I was always told that I was spoiled compared to some of the boys in my street, and especially compared to the boys at the top of the Ligoniel Road. I always thought that meant Catholic boys with their big families and their crammed houses, the same size as ours but packed with five or six of them to a bed, where they would sleep top to tail. They would run into their own house in the afternoon to dip dry bread in the sugar bowl that stayed in the cabinet in the front room, or they would nick a few spuds from the back of the potato lorry to roast in an open fire up the fields; they would beg food. 'You are spoiled rotten,' my mother would say in our back room with water running down the walls, 'and don't forget that.' Deprivation is, after all, always a relative concept. I knew that I had more toys than any of them but I didn't want to give the fort away.

I don't know where I got the hammer from; it must have been from the toolbox in the back room of my uncle's house. I must have had to search for it. It was a big heavy claw hammer. I hid it up inside my coat and said that I was going out. The fort was still there, just as I had left it, in the middle of the dump. No deprived child had got there yet. I sat down on the slope beside it. I suppose that it was almost like playing again. The first blow flattened two or three of the metal ramparts. The second removed one section of the metal steps. I sat on the dirty stones among the piles of rubbish and hammered away. I wasn't emotional about what I was doing. It was a cold act. I was just determined that no child, no matter how deprived or how needy or how hungry, would get my fort where my grandfather had fought for the British Empire, where Davy Crockett, whose father came from County Londonderry, had held out against the Mexicans at the Alamo, where my dreams of lands far away from cold damp mill houses that turned everything to rust had been nurtured.

I was obviously engrossed in my little frenzy of destruction because what I remember next is my father and uncle standing over me. They must have wondered where I had got to. My father looked almost puzzled, perhaps a little hurt that he had a son who could be like this. I looked up at them. I felt ashamed and embarrassed. I needed to explain my actions, to justify myself. I remember what I said quite clearly. 'It's dangerous,' I said. I remember those very words just coming out. 'Those sharp metal ends, they could hurt somebody. You can't just leave it here. Somebody might cut himself on it. I was just making it safe for them.'

I was led away by my father and my uncle, who didn't say anything or even look at each other. 'Let's just leave it here the way it is,' said my father eventually.

'But it's ruined now,' I said. 'It's ruined.' I was crying by now, sniffing loudly, wiping my nose on my sleeve. I remember looking down at the trail of smeared, green, thick mucus along my sleeve and thinking that there was just so much mucus. My voice as it sounded then is clear even now. It was a whining, imploring sort of crying that accompanied my excuses. But why I was crying I don't really know: perhaps it was being caught red-handed, the guilt of the whole thing, the fact that there was no way to hide my shame. Or perhaps it was just my way of showing them that I was still a child, who needed to dream, whose time had not come to leave these particular things behind.

This is a memory from my childhood but a memory that defines some of the more negative aspects of my own enduring character. I become attached to my possessions, they are part of me, I cannot give them away, I cannot recycle, I cannot hand them over. But how do we break the emotional bond with material objects? And how do we stop materialism feeding into concepts of self-identity and self-worth? How do we stop people feeling insecure when they cannot define themselves in any other way except through purchase or display? If we are going to do something about global warming then we will need to change many deeply ingrained habits. Some of this will not be easy. I know: believe me, I know.

PART III

Notes on dissociation

10

In two minds

This book was always going to be a journey, a journey in uncharted territory, maybe even a little stop–start at times, but I felt that some slow progress was being made. I now knew certain things. I knew that attitudes were generally positive towards low-carbon-footprint products and that people were, inside, relatively green (in some domains at least), but I also knew that people didn't really pick up on carbon footprint information in the time it takes to make a supermarket decision. The idea behind carbon labelling was (in principle) a good one, an empowering one. It allowed ordinary people to act in accordance with their underlying beliefs, but the actual mechanics of carbon labelling – how the information was represented, what kind of icons were used, what kind of numerical information was included, how the label looked – all needed a little bit more thought. Even those with the right underlying attitudes, the majority of the respondents, weren't attending to the information in that 5-second or 10-second envelope that is critical to the purchasing decision in the supermarket. And perhaps people didn't really care quite enough to find the carbon label icon in the necessary time frame. Their attention was not automatically drawn to it.

The research so far had also thrown up something else that was quite interesting. It allowed me to identify a certain group of people, between 10 and 20 per cent in the original sample, in which the two types of attitude (explicit and implicit) did not match, and it is to these individuals that I now turn. When I started thinking about this group a few months ago when I first saw the results, I called them 'the green fakers'. However, I grew to dislike this term; it is unkind and unfair, unfair to them and unfair to me because when I asked Laura to run me through the experiment I discovered that I was one of them. In order to understand these people (and myself) I needed to understand a bit more about how these two attitudes are actually represented in the human brain, and in order to do this, I needed to consider in much more detail the limitations of the research that I had done so far on underlying attitudes. I suspected that this held the key to a lot of the important issues.

The research so far on attitudes was, of course, exploratory and by necessity purely descriptive. It didn't really consider how the explicit and implicit attitudes were related within the individual. Were they significantly correlated or were they discrepant? Could there, in fact, be some level of 'dissociation' between them? ('Dissociation' is one of those psychological terms that seems so complex but at its heart is very simple – it just means 'being separate' or 'lack of connection', although I will discuss in a moment how psychologists use this term to refer to different states.) And if there are marked discrepancies within individuals, what are the psychological consequences for ordinary people (and me) of attempting to understand and explain their everyday actions concerning green choices in supermarket shopping? Furthermore, if implicit attitudes are largely unconscious, unlike explicit attitudes, can we rethink the role of unconscious impulses in everyday behaviour? Could we, in fact, look for evidence of the unconscious breaking through, perhaps as Freud (1901/1975) had done over a century earlier with respect to slips of the tongue? Any attempt to answer these questions will require a much fuller description of the relationship between implicit and explicit attitudes and a consideration of the new research into how thinking and speaking are represented and generated by the human brain. Interestingly, this will involve not just the analysis of speech itself (as Freud had done), but the analysis of human communicative behaviour more generally, because current research in psychology suggests that speech and the accompanying hand movements must be considered *together* to understand more fully speakers' underlying thought processes.

But first the relationship between implicit and explicit attitudes. There is now a *major* controversy within psychology about the relationship between these two constructs. Many psychologists maintain that the representations underlying explicit and implicit attitudes are, in fact, dissociated. According to Greenwald and Nosek (2008: 65):

> This reference to *dissociation* implies the existence of distinct structural representations underlying distinguishable classes of attitude manifestations. In psychology, appeals to dissociation range from the mundane to the exotic. At the mundane end, the dissociation label may be attached to the simple absence or weakness of correlation between presumably related measures. At the exotic end, dissociation may be understood as a split in consciousness, such as mutually unaware person systems occupying the same brain. (emphasis in original)

So how exotic is this dissociation in attitudes likely to be? There are a number of critical lines of evidence here. Nosek and Hansen (2008) report that in a meta-analysis of 81 studies the IAT was only moderately correlated with self-reported attitudes ($r = 0.24$; Hofmann, Gawronski, Gschwendner, Le and Schmitt 2005) and, in a study of fifty attitude domains, Nosek (2005) found that the strength of the correlation between the IAT and self-reported attitudes varied from near zero for some attitude domains (e.g. attitudes to thin and fat) to approximately 0.70 in other domains (e.g. pro-choice/pro-life attitudes). So in Greenwald and Nosek's (2008) terms there is clearly at least a mundane dissociation between the two constructs. But, in addition, there is the finding that other variables (for example, chronological age) can have a well-defined relationship with one of the measures (say, explicit attitude) but no relation with the other (say, implicit attitude). This has been found for things like explicit and implicit age preference. Here, there is a significant correlation between the age of the participant and age preference in the case of the explicit measure but no significant correlation in the case of the implicit measure.

Of course, these types of finding of possible attitudinal dissociation do not point to a single interpretation of the data, and Greenwald and Nosek (2008) have argued that they are, in fact, compatible with three different interpretations. First, the 'single-representation hypothesis', which maintains that the appearance of a dissociation is really just an illusion and all that is happening is that in explicit self-report measures, participants have the opportunity to modify their real response in their explicit reporting of their attitude. The second 'dual-representation hypothesis' is the real dissociation claim, and maintains that implicit and explicit measures of attitude have structurally distinct mental representations of attitudes and they are genuinely dissociated (see Chaiken and Trope 1999; Wilson, Lindsey and Schooler 2000). This hypothesis seems to be favoured by many in this area. Thus:

> Abundant theory, and some evidence, point to the human mind being divided into two largely independent subsystems: first, a familiar foreground, where processing is conscious, controlled, intentional, reflective and slow, but where learning occurs rapidly; and second, a hidden background, where processing is unconscious, automatic, unintended, impulsive and fast, but where learning occurs gradually . . . This dual-process model is appealingly neat. It recalls Freud's model of the mind, but with the inner sex maniac replaced by a dull but efficient zombie.
> (Gregg 2008: 764)

In this hypothesis the implicit attitudes operate automatically and unconsciously (and have unconscious representations) while the explicit attitudes operate consciously and with deliberate thought (and have an underlying representation which is quite different).

The third 'person vs culture hypothesis' is that explicit measures capture the attitudes operating within a person while implicit attitudes represent the more general influence of what is known about a particular thing in a particular culture. Nosek and Hansen (2008) argue, however, that the evidence from variations in the level of correlation in the IAT and self-reported attitude across individuals largely rules out this third hypothesis, and that the IAT does not merely reflect the evaluative judgement of the culture as a whole. Their conclusion is that the data demonstrate that the IAT is an individual difference measure and is associated with individual-level thoughts, feelings, and actions.

But that still leaves two major hypotheses: one (the single-representation hypothesis) is that there is one underlying representation for both implicit and explicit attitudes; the second is that implicit attitudes have their roots in the unconscious (dual-representation hypothesis) and explicit attitudes have their roots elsewhere. The first hypothesis might suggest that when we report our attitude we may be aware of what we really feel but we modify our verbal response particularly in sensitive domains (like race, attitudes to obesity and attitudes to green issues which we might like to exaggerate to make ourselves look good). The second hypothesis suggests that the implicit attitude is grounded in the unconscious; we are unaware of this attitude and may well be puzzled by the slowness of our reaction times and our high error rate when we sit in front of the computer screen as we complete the IAT.

This is clearly a major issue for social psychology and for those wishing to promote behavioural change in many core areas. For example, when we find using self-report measures that people seem to be pro-low carbon in explicit attitude (but not in implicit attitude), how should we interpret this result: as merely a social desirability effect or as something more profound? This could be extremely important from the point of view of behavioural change. One reason for saying this is that there are clear practical implications when the two attitudes align. As Greenwald and Nosek (2008) pointed out, 'When self-report and IAT measures were highly correlated with each other – a circumstance occurring especially in domains of political and consumer attitudes – both types of measures were more strongly correlated with behaviour than when implicit–explicit correlations were low' (2008: 78). In other words, once we know something about the nature of the correlation between these two measures, we will have a much better understanding of the likelihood of predicting behaviour from either or both of the measures. This is of major significance when it comes to climate change and how to tackle it.

So, I returned to the original data to take a fresh look at the scores and I carried out a batch of statistical tests, and in particular a series of correlations to work out the relationship between the various measures. A Pearson product–moment correlation coefficient revealed that there was no significant correlation between any of the measures of explicit attitude and the IAT measure (for the Likert scores and D scores $r = 0.008$, for the thermometer difference scores and D scores $r = -0.057$). A Pearson product–moment correlation coefficient, however, showed that there was a strong positive relationship between the various measures of explicit attitude (namely, the Likert and thermometer difference scores, $r = 0.560$) as shown in Figure 10.1.

Further comparisons using a Spearman's rank-order correlation coefficient suggested that while there was no relationship between age and D scores ($\psi = 0.56$) as shown in Figure 10.2, there was a positive relationship between age and Likert scores ($\psi = 0.214$) as shown in Figure 10.3

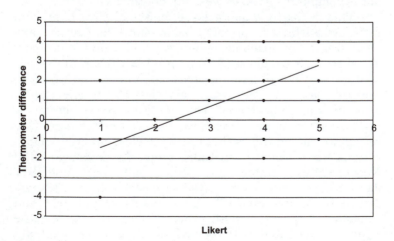

Figure 10.1 **Likert and thermometer difference correlations.**

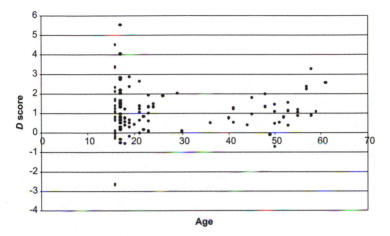

Figure 10.2 Age and *D* score correlations.

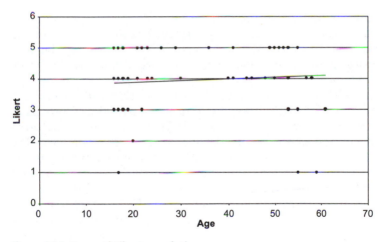

Figure 10.3 Age and Likert correlations.

and between age and thermometer difference scores (ψ = 0.197) as shown in Figure 10.4.

In other words, the measures of implicit and explicit attitudes could well be dissociated because firstly there is no significant correlation between them, and secondly one additional variable (chronological age) correlates with one

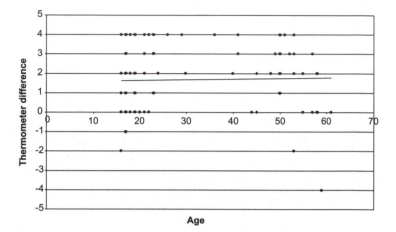

Figure 10.4 **Age and thermometer difference correlations.**

(the explicit measure) but not the other (the implicit measure). Of course, this latter finding is interesting because it suggests that as people get older they become more concerned about explicitly reporting their green credentials.

The fact that there was no significant correlation between the implicit and explicit measures also allows us to identify different sets of individuals with differing patterns of implicit and explicit attitudes. Figure 10.5 compares the Likert and D score results. While the majority of participants showed some degree of convergence between implicit and explicit attitudes with D scores > 0.8 and Likert scores of 4 and 5 (outlined by a dashed line in Figure 10.5), there were 13 participants in this first sample of 100 who, despite saying explicitly that they were very pro-low carbon with Likert scores of 5, had implicit attitudes that were not as positive, with D scores < 0.8 (outlined by a bold dashed line in Figure 10.5).

Figure 10.6 compares the thermometer difference and D score results. Again, the majority of participants displayed convergent attitudes, expressing a preference for low-carbon-footprint products at both the implicit and explicit levels (outlined by a dashed line on the right-hand side of

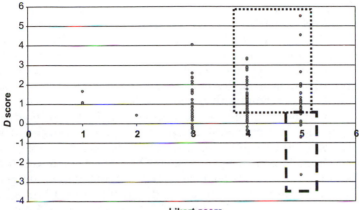

Figure 10.5 Likert and *D* score comparisons.

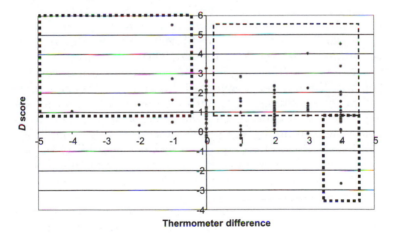

Figure 10.6 Thermometer difference and *D* score comparisons.

Figure 10.6). However, there were two sets of participants that showed attitudinal divergence at both extremes. Overall, twelve participants (a similar number to those who displayed divergent attitudes using the Likert scale) explicitly stated that they were pro-low carbon with thermometer difference scores of 4, but their implicit attitudes were much less positive, with D scores < 0.8 (outlined by a bold dashed line on the right-hand side of Figure 10.6). Interestingly, a second set consisting of five participants (outlined in bold to the left of Figure 10.6) expressed explicit attitudes that demonstrated a preference for high carbon (using the criterion of < 0 for thermometer difference scores); however, implicitly their attitudes appeared to be much more positive towards low carbon (with D scores > 0.8).

To recap, although the explicit measures of attitude (the feeling thermometer and Likert scale) were significantly correlated, the explicit and implicit measures were not correlated. This discrepancy seems to reflect some kind of 'dissociation', and has been reported previously in a number of other domains (see Banaji and Hardin 1996; Blair and Banaji 1996; Bosson, Swann and Pennebaker 2000; Devine 1989; Fazio, Jackson, Dunton and Williams 1995; Greenwald, McGhee and Schwartz 1998), although in other domains measures of implicit and explicit attitudes appear to be positively related (see Greenwald and Nosek 2001; Hofmann et al. 2005; Nosek and Banaji 2002; Nosek, Banaji and Greenwald 2002; Nosek and Smyth 2005).

Nosek (2007) has argued that 'measurement innovations [such as the IAT] have spawned dual-process theories that, among other things, distinguish between the mind as we experience it (explicit), and the mind as it operates automatically, unintentionally, or unconsciously (implicit)' (2007: 184). So we have here the distinct possibility of two largely independent subsystems in the human mind, one that is familiar and one that is not. (Whether we have *any* 'conscious' awareness at all of our implicit thinking, and whether the implicit process is always truly unconscious or whether we have some inkling of the underlying evaluation, remain to be properly investigated. The fact that something cannot be consciously controlled and manipulated does not of course mean that it resides purely and totally in the unconscious.) But how does this divergence between implicit and explicit attitude manifest itself within the individual, and does it have any effect on any aspects of observable behaviour? After all, a hundred years ago or so Freud showed how unconscious (and repressed) thoughts could find articulation through the medium of everyday speech in the form of slips of the tongue. And how might this dissociation impact on people's willingness or ability to actually do something about climate change? These are potentially important questions from both a theoretical and a practical point of view. It surprised me that nobody until now had attempted to answer them.

11

Speech and revealing movement

In order to make a first attempt at an answer here, we have to accept a major challenge and start thinking afresh about the very nature of everyday communication in which people express their underlying thoughts and ideas. After all, if we want to see the unconscious at work we must know where to look. When human beings talk, you will have noticed that they make many bodily movements, but in particular they make frequent (and largely unconscious) movements of the hands and arms. They do this in every possible situation – in face-to-face communication, on the telephone, even when the hands are below a desk and thus out of sight of their interlocutor (I have many recordings of these and similar occurrences). It is as if human beings are neurologically programmed to make these movements while they talk, and these visible movements would seem to be (in evolutionary terms) a good deal more primitive than speech itself, with language evolving on the back of them. These gestures are imagistic in form and closely integrated in time with the speech itself. They are called 'iconic gestures' because of their mode of representation. Words have an arbitrary relationship with the things they represent (and thus are 'non-iconic'). Why do we call a particular object a 'shoe' or that large four-legged creature a 'horse'? They could just as well be called something completely different (and, of course, they are called something completely different in other languages). But the unconscious gestural movements that we generate when we talk do not have this arbitrary relationship with the thing they are representing: their imagistic form somehow captures certain aspects of the thing that they are representing (hence they are called 'iconic') and there is a good deal of cross-cultural similarity in their actual form (see Beattie 2003, Chapter 6).

Just visualise someone speaking, when they are fully engrossed in what they are saying, in order to understand the essential connections here between speech and hand movement. Below is a speaker who seems pretty engrossed in what he is saying. It is taken from a clip on the internet. It is Steve Ballmer, the CEO of Microsoft, and he is talking in an interview about future development of the PC and other 'intelligent edge devices':

Steve: The PC is an important part of the [overall ecosystem] that people are using . . .

Iconic: Hands are close together, forming a sphere.

Steve: I think there's gonna be [two places of innovation] for development over the next few years.

Iconic: Hands are closed into fists, a foot apart.

Figure 11.1 **Examples of the iconic gestures that accompany talk.**

Just look at the elaborate hand movements, drawing out images in the space in front of his body. These imagistic gestures do not merely 'illustrate' the content of the speech; rather they are a core part of the underlying message. Steve Ballmer does not say what he intends to say and then try to make it clearer with a gestural illustration, after a brief pause. Rather he uses speech and movement simultaneously; the movements and the words both derive from the same underlying mental representation at exactly the same time (actually the beginning of the gestural movement slightly precedes the speech so that the hands can be in exactly the

Steve: I think that [people got very excited], appropriately, about the internet, html and browsers.
Iconic: Hands start off in front of the body and make a fast sweeping motion to the right of the body.
Figure 11.2 **Further examples of the iconic gestures that accompany talk.**

right position to make the critical movement at the right time). That way the gestures are perfectly timed with the speech, and together with the speech they form a complete whole. The two systems are perfectly coordinated.

The gestures also seem to be generated without any conscious awareness. When people are talking they will know that their hands have just done something, that they have made some movement, but if you ask them to make the same gesture again they find it very difficult to do this, or if you ask them what exactly the gesture was communicating, they will say 'I have no idea.' Many gestures contain a complex of different images: when asked to repeat the movement, speakers may make a stab at repeating one of these. They will know where in front of their body they made the movement but usually this is about the only thing that they will get exactly right (unlike speech itself, which we are pretty accurate at repeating and reproducing).

It may seem strange, considering the emphasis now placed by some psychologists on the imagistic gestures that speakers generate unconsciously while they talk, that until quite recently the historical view regarding such gestures was that they constituted a system very much secondary to speech, only really useful in noisy or difficult environments and not very accurate or precise (see Kendon 2004 for an overview).

My favourite story about the lack of precision and base inaccuracy of gestural communication involves the infamous Charge of the Light Brigade in 1854 during the Crimean War. The Heavy Brigade of cavalry led by General James Scarlett had made an uphill charge against the Russians in the plain of Balaklava. Five hundred British horsemen were pitted against three thousand Cossacks, but the Heavies prevailed after fierce fighting, and the Russians began to retreat. Lord Raglan, the Commanding Officer of the British forces, watched from the vantage point of the Sapoune Heights as the Russians started to escape back up the North Valley, pulling their cannon with them. Raglan sent a series of orders to Lord Lucan, the Commander of the Light Brigade, telling him to attack the guns that the Russians were attempting to pull away from either side of the valley. But the only guns that Lucan could see were the heavy gun emplacements a mile down the valley, and it would have been suicidal to attack these (see the excellent description in Cummins 2008: 199).

> An hour and a half passed with Lucan attempting to get clarification from Raglan and Raglan becoming more and more impatient. Finally, fed up, Raglan dictated a note that read in part: 'Lord Raglan wishes the cavalry to advance rapidly to the front – follow the enemy and try to prevent the enemy from carrying away the guns.' The note was given to a captain Lewis Edward Nolan, one of the finest young cavalry officers in the British army ... Riding pell-mell down the steep cliff to the valley, he arrived in front of Lucan and impatiently delivered the message to attack the Russian guns.
>
> Angered by Nolan's arrogance, Lucan replied: 'Attack, sir? Attack what? What guns, sir?' And Nolan, instead of pointing at the guns along the Causeway, flung his arm in the direction of the Russian emplacements a mile and a quarter down the valley: 'There, my lord, is your enemy, there are your guns!' (Cummins 2008: 199)

It was Lucan's brother-in-law, James Thomas Brudenell, the seventh Earl of Cardigan, who received the order from

Lord Lucan to lead the Light Brigade into the Valley of Death (an event, of course, immortalised in Alfred Lord Tennyson's famous poem 'The Charge of the Light Brigade'). Lucan was to follow with the Heavy Brigade. Cardigan was apparently heard to mutter, 'Here goes the last of the Brudenells' after he received his order. It was an extremely brave but foolhardy charge. Twenty minutes after the charge began only 195 out of 673 cavalrymen survived, and the whole thing would not have occurred without Captain Nolan's infamous gesture.

The communication between Lords Raglan, Lucan and Cardigan was difficult for a number of reasons. They all had different perspectives on the valley: Raglan could see the whole thing unfold, Lucan and Cardigan could not see which guns Raglan was referring to. In addition, Lucan and Cardigan were feuding aristocrats (and feuding brothers-in-law who basically despised each other), so their relationship did not help the essential flow of the communication. So neither the physical nor what you might call the social context allowed for smooth communication here, and the whole communicational problem boiled down to what 'the guns' actually referred to. Nolan attempted to clarify 'the guns' with a gesture but the gesture was poorly formed, inaccurate and misconstrued (according to historical record). It was, thus, in many historians' eyes, a single gesture that led to the Charge of the Light Brigade and the loss of so much life (as well as producing one of the most enduring and iconic images of the stiff upper lip of the British aristocracy in action, with Lord Cardigan leading his men bravely to what he thought was certain death, although he himself did survive).

So the historical view is that gestures are a poor form of communication, perhaps necessary in noisy or difficult environments (factories, talking to foreigners in one's own language, heroic and tragic battles), but really only an add-on to speech, rather superfluous, and only really necessary when verbal language is itself stretched. The transformation of the way in which gestures are viewed began with the pioneering and influential work of David McNeill (1985). Through his careful and painstaking analyses, McNeill demonstrated that this view is simply wrong. He showed that these gestures are an integral component of everyday communication in every context imaginable, a core part of the process of representing and communicating ideas, and in many ways every bit as significant as speech itself. He showed that if you want to understand a speaker's underlying representation of an idea, then you need to hear the speech and see the gesture (the fact that our interpretation of the gestural information is done unconsciously is neither here nor there). He showed that it is only when the two channels are combined that the full message of the speaker is successfully conveyed (see also Beattie 2003).

Figure 11.3 shows a very simple example from McNeill (1992: 13) that demonstrates some of the basic principles underlying this everyday communication, which involves both speech and gesture. What this example from a cartoon narration reveals is that within the speech itself there are details of the 'action', 'the characters involved' and 'the concept of recurrence of a chase', yet there is no mention of any weapon being used in the pursuit. However, the

'she chases [him out again]'
Iconic: Hand gripping an object swings from left to right.
Figure 11.3 McNeill's example of a simple action sequence.
Source: McNeill, D. (1992) *Hand and Mind. What Gestures Reveal About Thought*. With permission by University of Chicago Press.

gesture used alongside the speech portrays the weapon being brandished and communicates effectively why one character is running from another. Of course, the speaker could have said 'she chases him out again with an umbrella' but didn't. It is as if the brain is sending the message about what needs to be communicated to the speech system and the gestural system simultaneously and, therefore, in order to understand the core message both the gesture and the sentence must be taken into account by the recipient of the message.

As McNeill (2000: 139) says, 'To exclude the gesture side, as has been traditional, is tantamount to ignoring half of the message out of the brain.' The critical point is that if the receiver attends only to the speech, they will be missing out on this additional information. While the speech and gesture in this example are obviously connected in terms of their semantic content, they are not identical. As such, this gesture is said to be 'complementary' to speech (McNeill and Duncan 2000).

Although speech and gesture may communicate, they operate very differently as semiotic systems. Figure 11.4 shows some pictures of a student talking about a table being raised towards the ceiling.

Once you start studying the sentence and the gesture, you can see some striking dissimilarities in how they function. Speech operates in a linear and segmented fashion, identifying what is being raised ('the table'), the action ('can be raised up') and the direction of the action ('towards the ceiling'), all sequentially in a linear and segmented way.

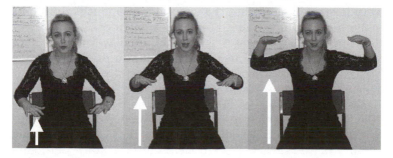

Figure 11.4 Gesture representing a table being raised towards the ceiling.

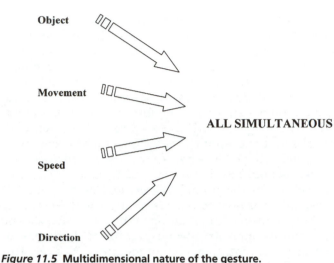

Figure 11.5 **Multidimensional nature of the gesture.**

The gesture is, however, multidimensional, representing the object, the movement, the speed and the direction simultaneously.

To understand speech we have to proceed in terms of bottom-up processing where we start with the individual words and interpret the meaning of words before trying to understand the sentence using both the word meanings and the syntax. Gesture operates much more in a top-down fashion. We need to understand that this particular gesture is representing a table moving upwards to be able to interpret that the hands wide apart are representing that it is a very large table and that the upward movement is telling us something about the speed of the movement of the table.

Speech also operates with individual words that have standards of form. If I use a word that you don't recognise, you can ask me whether it is a proper word and what it means. But these gestures don't have standards of form. They are not like sign languages of the deaf, nor are they like 'emblems', which are specifically coded gestural forms with a direct verbal translation (like the V sign meaning victory or the reverse V sign meaning f**k you). Both sign languages and emblems have standards of form: when

Figure 11.6 **A famous V-sign: an emblem with standards of form.**
Source: Bettmann-Corbis.

Margaret Thatcher famously gave a 'V for Victory' sign to a group of journalists after her 1979 General Election win, but with the back of the hand facing the journalists, they were in a position to query the exact message that she was sending (and apparently some of them did query it, but not with the great woman herself!). It could, of course, have been her unconscious desires breaking through into overt behaviour. She might have been telling the journalists exactly what she thought of them.

But the spontaneous imagistic gestures that accompany everyday talk do not have standards of form like this. You cannot stop someone mid-flow and ask them why they used a particular imagistic movement. With these movements human beings create meaning spontaneously and unconsciously for other human beings without the benefit of a dictionary, and they seem to do this effortlessly and extremely frequently.

Stemming from the initial research of McNeill, the role of gesture in communication is now being extensively studied, opening up a new field within psychology. Such research has consistently revealed that in order for the receiver to obtain the full meaning of an utterance, the two channels of communication need to be combined in order to form a more complete, overall representation of the message (Beattie and Shovelton 1999a, 1999b, 2001, 2002a, 2002b, 2006, 2009). So in one set of experiments we asked people to tell us what was going on in a set of cartoon stories and we filmed them telling these stories. We then played parts of their speech, with or without gestures, to another set of participants who were then quizzed about the original story ('Was Billy sliding to the left?'). We found that where experimental participants were played either extracts of speech on their own or extracts of speech accompanied by the iconic gestures (played back on a video-screen), there was a clear advantage to seeing the gestures in addition to hearing the speech in terms of gaining information about the original stories (even though our participants often reported that they didn't really *notice* the often small and insignificant gestures that occurred with the speech and weren't aware of trying to decode them in any way). From these studies it became clear not only that significant information is encoded in the gestures that accompany speech, but also that receivers are able to decode this information successfully and quickly and combine this information with that encoded in the speech itself, and they are able to do this without any conscious awareness of where they got the information from.

There are, of course, clear practical applications of this theoretical perspective. If significant semantic information is naturally split between the speech and the gestural channels, is it most effective to communicate information about (say) a commercial product using both channels, and how effective are these gestures at communicating aspects of the product compared to other types of images? With Heather Shovelton (Beattie and Shovelton 2005) I applied this theoretical perspective to television advertising (with the financial support of Carlton Television) to see whether gestures were more effective at communicating certain core properties of a product than the kinds of images that advertisers might traditionally use. We made some specific ads to promote an imaginary, but plausible, fruit drink (indeed, so plausible was it that it subsequently came onto the market!). In the ad two characters talked to camera about a new fruit drink 'F' ('with your five portions of fruit in one tiny little drink'). For one version of the ad, the speech–image advertisement, an advertising agency created its own images to convey three core properties of the product (*freshness* – juice sparkling on the fruit; *everyone is drinking it* – the *Stun* (*sic*) newspaper displaying the headline 'Phwoar! Everyone's drinking it'; *the size of the bottle* – the image of the actual bottle with respect to the hand), as shown in Figure 11.7.

For the other version of the ad, the speech–gesture advert, we included three iconic gestures that represented core aspects of the product, including: that the fruit was 'fresh' (hands together in front of chest, they move away from each other abruptly as fingers stretch and become wide apart), that 'everyone' was drinking it (right hand and arm move away from the body making a large sweeping movement) and the 'size' of bottle (hands move towards each other until they represent the size of the bottle), as shown in Figure 11.8. These gestures were selected on the basis of the fact that they were the kinds of unconscious movements that

Figure 11.7 Images representing 'freshness', 'everyone' and 'size'.

Figure 11.8 Gestures representing 'freshness', 'everyone' and 'size'.

people habitually make to represent these features when talking about these kinds of things. There tended to be a high degree of commonality in these representations across individuals (even without a dictionary of possible gestural representations to draw on).

We played these different versions of the ads to two independent groups of participants, and discovered that the use of speech and gesture together (including both the concrete 'iconic' gesture for 'size' and the more abstract 'metaphoric' gestures for 'freshness' and 'everyone') were more effective at communicating the core semantic features of the product than the speech and image version. So, it turns out that not only are these gestures highly informative, they are significantly more informative than other types of images that we might select (consciously and deliberately and with great creative thought) to represent core aspects of the product.

What is also interesting about the relationship between gestures and speech is that listeners interpret the gestures, and extract the critical information contained within them, without any apparent conscious awareness. Ask them after-wards how they managed to pick up the critical details when they are only contained within the accompanying ges-tural movements, and generally speaking they don't have a clue. They normally seem to assume that the speaker included that information in the speech itself, and only in the speech ('there was someone waving his hand about, but I didn't pay too much attention to that' is a fairly common response).

However, there are times in everyday conversation where the speech and gesture channels do not gel in the normal way but rather the two channels may appear to con-tradict one another; this has become known as a 'gesture–speech mismatch' (Church and Goldin-Meadow 1986). Mismatches may occur for a variety of reasons. I have argued elsewhere that they occur when a speaker is trying to conceal critical information from a listener (see Beattie 2003). The basic idea here is that speakers control and edit their speech (because it is a conscious and reflective medium) when they are trying to tell a lie, but since they have less conscious control over their imagistic gesture, the spontaneous imagistic gesture emerges untarnished and reveals their underlying thought regardless. Hence, on occasion we find gestures and speech that do not match because the speech has been changed; the gesture has not.

Here are some examples of gesture–speech mismatches from various television programmes that I have worked on (Beattie, *Big Brother* 2000–2009, Channel 4, UK; Beattie, *The Body Politic on News at Ten Thirty*, 2005, ITV, UK) with a possible explanation of why they occurred in the first place. The first example comes from one of the housemates, Adele, in the UK's *Big Brother* Series 3. Here Adele is asked by the anonymous voice of 'Big Brother' who she thinks will be evicted by the public vote that evening. She uses a list structure (see Jefferson 1990) to give the order of who she thinks will go, starting with a contestant called Jade (soon to become very famous but who has since tragically died of cancer), then herself, then Jonny, then Kate. In other words, in her speech she is saying that she herself is very likely to be evicted from the Big Brother house (she is saying, in effect, that she thinks she is in second position to Jade, and therefore has a high probability of going). But her gesture seems to be communicating something quite different here, as described in Figure 11.9.

One hand (the left hand) represents Jade, the right hand represents herself and the other two contestants (the square brackets indicate the start and end points of the meaningful part of the gesture, the so-called stroke phase). The gesture shows that she actually thinks that Jade is by far the most likely to be evicted from the house and that the other three (represented by a different hand, indicating considerable psychological distance between Jade going and the other three) are all safe. This interpretation is supported by the fact that when Adele was actually evicted that very evening, she was genuinely surprised by her eviction and the public vote. This is a gesture–speech mismatch because the speech seems to imply that she thought that she was likely to be evicted; the gesture does not show this.

The second example of a gesture–speech mismatch comes from the UK's *Celebrity Big Brother* Series 2, and involves a well-known comedian called Les Dennis. Here, Les was the only housemate in a position to nominate for the forthcoming eviction (because he got zero on a Big Brother quiz) and he was explaining to Big Brother why the situation that he was in was so difficult for him. It's important to remember that he knew that he wouldn't have looked so good to the great British public if he had said it was going to be

---

Adele: [So Jade, then me, then Jonny, then Kate], I think that's the order.

Iconic: Hands and arms are wide apart and resting on the arms of a chair. Left hand rises slightly with index finger pointing forwards as she says 'Jade'. Right hand then rises slightly as she says 'then me', index finger points forwards, finger moves slowly to the right and as it does so it makes three slight up-and-down movements.

Figure 11.9 Adele's gestures revealing what she is actually thinking.

easy to make the nominations ('what a heartless man!'). This was his essential dilemma. So while Les was saying in his speech that the housemates were all really close (thus making it very difficult to nominate any of them), his gestures would suggest a very different interpretation. You would expect that if Les did, in fact, think the housemates were really close then his hands would have been much closer together during the generation of the gesture, reflecting this degree of closeness (because people do use positioning in the gestural space in front of the body in a consistent and meaningful way). However, in the critical gesture the hands were much further apart than one would expect, indicating that the housemates were not as close as Les was suggesting in his speech. Indeed the gesture was away from the other hand and away from the body. The hands were, in fact, signalling a significant psychological or emotional distance between the housemates, as described in Figure 11.10. Talking to Les, after the show was over, suggested that this interpretation was correct.

I have also identified mismatches in the talk of many politicians, including speeches of Tony Blair (then Prime Minister of the UK). In one particular speech (at the launch of the Labour manifesto in the run-up to the General Election in 2005), Blair was talking about possible rises in

---

Les: We [are all six of us, very, very, close]

Metaphoric: Left hand is in front of left shoulder, palm is pointing forwards and fingers are straight and apart. Hand moves quickly to the left away from the body and then moves quickly back to its position in front of shoulder. This whole movement is repeated twice. The first half of the movement is then produced for a third time and the hand now remains away from the body.

[really close]

Metaphoric: Hands are wide apart, palms point towards each other. Hands move rapidly towards each other to an area in front of stomach but hands don't touch – they stop when they are about six inches apart.

Figure 11.10 Les' gestures indicate a psychological distance between the housemates.

National Insurance contributions to fund developments in the NHS in the next Parliament, if Labour were to be returned to power. At the beginning of the sentence, Blair sweeps his left hand from the left side of his body to the centre position of his body as he talks about rises in National Insurance contributions in the last Parliament. He then says that these rises will continue to fund developments in the NHS through the next Parliament. One would expect the gesture to continue moving across the body, signalling this continuation. However, instead of continuing to move across the body, the hand stops halfway across the body when he says 'last Parliament' and rather unnaturally sticks there, as shown in Figure 11.11.

The abrupt halt of the accompanying gesture may be interpreted as an indication that Tony Blair did not genuinely feel that those past rises would be enough to continue to fund NHS development in the succeeding period of Parliament.

I argue that in each of these examples people may be revealing what they are *actually* thinking (see Beattie 2003). While the speech can be consciously edited and controlled, the gestures are difficult, if not impossible, to edit or control in *real* time, and so the true thoughts and feelings of the speaker may become manifest in the gesture.

Tony Blair: '[Rises in National Insurance contributions funded development in the NHS right through the last Parliament] and will continue to fund them through the next Parliament.

Figure 11.11 Tony Blair's gesture–speech mismatch.

Source: ITN Source.

Although researchers have examined gesture–speech mismatches in situations like this, no one, thus far, has looked for gesture–speech mismatches where there is a clash between a person's implicit and explicit attitudes. But we might well expect those whose implicit and explicit attitudes diverge to display some evidence of this in gesture–speech mismatches. In contrast, people whose attitudes converge should show a higher level of matching speech and gestures (although we might find some evidence of mismatches here which we will have to consider in detail: mismatches may, after all, occur for a variety of other reasons). Gesture–speech mismatches, thus, could potentially allow us to pinpoint individuals whose underlying attitudes are not conducive to green behaviour (regardless of what they actually say). This could prove extremely useful in the future; and also very embarrassing to the likes of me, who might well fit into this group of non-congruent fakers.

12

In search of the green fakers (in search of myself)

Thus, from a contemporary psychological perspective, talk is seen as a complex multichannel activity which involves the expression of thoughts and ideas through both language and expressive movement, and particularly through the expressive movements of the hands and arms (see McNeill 1992, 2000; Beattie 2003). These expressive movements are imagistic and iconic in form and closely temporally integrated with the speech itself. Ideas are jointly expressed through the speech and the movement (Beattie and Shovelton 1999a, 1999b, 2001, 2005) and this has led David McNeill to the startling conclusion that 'To exclude the gesture side, as has been traditional, is tantamount to ignoring half of the message out of the brain' (2000: 139).

It is important to point out that there are different conceptual models of how this whole process of gestures and speech cooperating to communicate meaning actually works. McNeill (2005) proposed a psychological model based on the rather complex concept of the 'growth point' – the minimal unit of an imagery–language dialectic.

> A growth point is a package that has both linguistic categorical and imagistic components, and it has these components irreducibly. It is a minimal unit in the Vygotskian [Vygotsky 1986] sense, the smallest package that retains the property of being a whole; in this case the imagery–language whole that we see in synchronized combinations of co-expressive speech and gestures. (McNeill 2005: 105)

In McNeill's model, the construction of meaning and talk is 'a dynamic, continuously updated process in which new fields of oppositions [his terminology for a particular understanding of context] are formed and new GPs [growth points] or psychological predicates are differentiated in ongoing cycles of thinking and speaking' (McNeill 2005: 107). McNeill shows how this model can explain the form and timing of the gestural movements that accompany speech. In an example where a participant retells a cartoon story, the concept of the growth point is illustrated when someone says verbally 'drops it [a bowling ball] down the drainpipe'. The accompanying gesture has a distinctive shape and is not the gesture shown in the original cartoon. McNeill's conclusion is that 'The gesture and sentence . . . reflected the speaker's conceptualizing of the cartoon as much as the objective cartoon content' (McNeill 2005: 121). McNeill's (2005) model in which he makes speech and gesture absolutely integral to the process of meaning generation gives us a new way of analysing talk to glimpse the conceptualisation process of utterances in real time.

Others have recognised that gesture is an integral aspect of everyday communication but have not necessarily subscribed to the growth point theory. One other influential model is the Information Packaging Hypothesis (IPH) of Kita (2000), discussed extensively by McNeill. The IPH considers speech and gesture to be independent cognitive streams in speech, running simultaneously. The IPH is more of a modular conception of speech and gesture with the two modules as 'separate' intertwining streams (in McNeill's words). McNeill says that the imagery in the gesture is categorised linguistically, whereas in the IPH gesture is viewed as visual thinking. The IPH requires an interface for the imagery and gesture modules for the exchange of relevant information. McNeill's growth point model does not have such an interface because here gesture and language combine dialectically (see McNeill 2005: 132).

Thus, there are a number of significantly different theoretical interpretations of the exact relationship between gesture and speech, but they all agree on a number of theoretical points, mainly that gesture is an essential component of speaking and that communication between conversational partners depends critically on this component (see also Beattie and Shovelton 1999a, 1999b, 2001, 2005). The other thing they agree on is that our very conception of the nature of human communication has changed in the past few years.

There is one feature of gestural communication in particular that might be extremely relevant to our current considerations, and that is how unconscious this process of generating gestures actually is. Speakers, as I have suggested, may have good conscious awareness of how their speech is unfolding in real time but they seem to be much less aware of the exact form and timing of their gestural movements. This has been emphasised by a number of leading researchers in this area. Thus Cienki and Müller (2008) wrote that 'Gestures are less monitored than speech and they are to a great extent unconscious. Speakers are often unaware that they are gesturing at all' (2008: 94). Danesi (1999) argued that 'when people speak they gesture unconsciously, literally "drawing" the concepts they are conveying orally' (1999: 35). Nelson (2007) stated that:

> gesture is a parallel component of human verbal communication, sometimes used unconsciously to accompany the message conveyed by words (Goldin-Meadow 1997) . . . [gestures] are often acquired and used without conscious intent. To the extent this is the case, it verifies the continuing existence of a mode of unconscious meaning unconsciously expressed. (1997: 96)

The possibility of using this mode of unconscious meaning unconsciously expressed to gain insights into the implicit aspects of underlying attitudes is an intriguing one. Could we analyse the gestures and speech in detail, specifically focusing on individual gestures and speech that do not match in terms of meaning for a possible insight into implicit–explicit discrepancies? Could we find evidence of dissociation between implicit and explicit attitudes through a micro-analysis of the everyday behaviour of people simply talking about their views? Further, if the gestures are a

Figure 12.1 Professor Geoff Beattie giving a talk in London. Even though he has studied gesture for many years, he was still unaware of what this particular gesture meant (it is much more than a pointing gesture, by the way).

mode of unconscious meaning unconsciously expressed, can we 'read' the implicit attitudes of the speaker in this mode of representation, even when the conscious verbal channel says something quite different?

Of course, a parallel sort of enquiry was started over a century earlier by Freud (1901/1975) in his analysis of slips of the tongue. Since the first descriptions of such slips (Meringer and Mayer 1895) there has been a widespread difference of opinion on what kinds of mechanisms are required to explain them. Wundt (1900) attempted to explain them through the 'contact effect of sounds' or what subsequent generations of linguists and psychologists might call psycholinguistic mechanisms – the processes and rules that generate speech production (see Ellis and Beattie 1986). But Freud was adamant that:

> Among slips of the tongue that I have collected myself, I can find hardly one in which I should be obliged to trace the disturbance of speech simply and solely to what Wundt (1900: 392) calls 'the contact effect of sounds'. I almost invariably discover a disturbing influence in addition which comes from something *outside* the intended utterance; and the disturbing element is either a single thought that has remained unconscious, which manifests itself in the slip of the tongue and which can often be brought to consciousness only by means of searching analysis, or it is a more general physical motive force which is directed against the entire utterance. (1901/1975: 103)

His carefully chosen examples seem to support his thesis. There may be a phonetic similarity between the origin and target of the slip, but there may well be some additional evidence of the unconscious breaking through into the conscious medium of speech.

Thus, according to Freud:

> A slip of the tongue had a similar mechanism in the case of another woman patient, whose memory failed her in the middle of reproducing a long-lost recollection of childhood. Her memory would not tell her what part of her body had been grasped by a prying and lascivious hand. Immediately afterwards she called on a friend with whom she discussed summer residences. When she was asked where her cottage at M. was situated she answered: 'on the *Berglende* [hill-thigh]' instead of *Berglehne* [hill-side].
>
> When I asked another woman patient at the end of the session how her uncle was, she answered: 'I don't know, nowadays I only see him *in flagranti*.' Next day she began: 'I am really ashamed of myself for having given you such a stupid answer. You must of course have thought me a very uneducated person who is always getting foreign

words mixed up. I meant to say: *en passant*.' We did not as yet know the source of the foreign phrase which she had wrongly applied. In the same session, however, while continuing the previous day's topic, she brought up a reminiscence in which the chief role was played by being caught *in flagranti*. The slip of the tongue of the day before had therefore anticipated the memory which at the time had not yet become conscious. (1901/1975: 105–106)

But imagine if Freud had had video recordings to work with and the new model of how thoughts are expressed through both speech and movement. What might a similar analysis of gesture, capable of generating its meaning well below the radar of consciousness, have revealed? That is the question we tackle here. Furthermore, we can be more targeted in our quest. With Freud, all we have is the observed behaviour; the rest is inference. Here we will have the behaviour – direct concrete evidence of gestures and speech that match or fail to match – but, in addition, we will have our independent measures of what their implicit and explicit attitudes actually are. This should help us focus our search and our interpretation in a much more systematic way.

From the IAT and the explicit attitude measures, we were able to identify two distinct sets of participants – one set had very positive attitudes to low carbon products on both the explicit and implicit measures ($n = 10$); the other had the greatest explicit and implicit attitudinal clash ($n = 10$). This sample of just twenty people was then contacted and asked to take part in an interview. Each person was filmed as they responded to a series of general questions regarding environmental issues and their own particular lifestyles. This included questions about what environmental behaviour they engaged in, what they knew about carbon labelling, whether they thought that carbon labelling could make a difference to global warming and whether they felt they themselves could make any difference to climate change. The aim of the interview was to get the interviewee to talk openly about environmental topics in an informal situation where they were hopefully relaxed enough to feel comfortable (and gesture freely). Afterwards the iconic and metaphoric gestures were identified and the accompanying speech was transcribed in detail.

The transcripts of gestures and speech use the following symbols.

[ ]   Square brackets indicate the beginning and end of the gesture.

[ ]   Square brackets in bold highlight the gestural movement that is of particular significance.

:   Colons are used to represent pauses in speech, where the number of colons indicates the length of the pause, e.g. :::: would indicate a very long pause.

(1)   A subscript number in parentheses indicates the sequence in which the gestures occurred

Figures 12.2 to 12.4 show the average number of gestures produced by the participants whose explicit and implicit attitudes either converge or diverge, the average time each

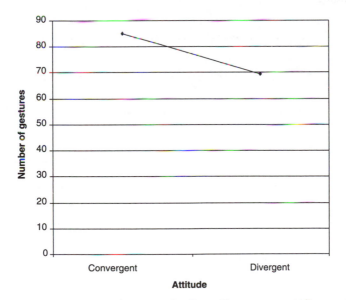

Figure 12.2 **How gesture frequency is affected by convergent/divergent attitudes.**

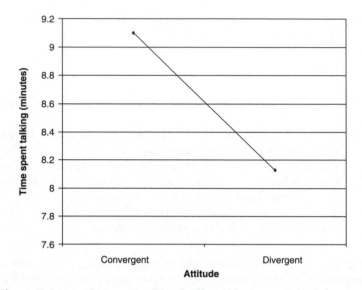

Figure 12.3 How time spent talking is affected by convergent/divergent attitudes.

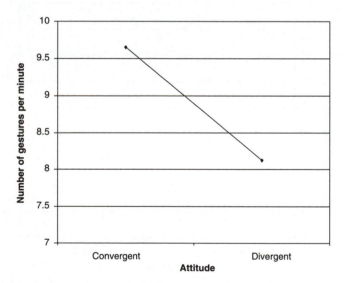

Figure 12.4 How gesture rate is affected by convergent/divergent attitudes.

Table 12.1 **Comparisons of the average number of gestures produced, average time spent talking and average number of gestures per minute for the convergent and divergent groups**

	Average no. of gestures	Average time spent talking (minutes)	Average number of gestures per minute
Convergent	85.10	9.10	9.65
Divergent	69.40	8.13	8.13

group spent talking and the average number of gestures produced per minute. Table 12.1 compares the data in these figures.

It would seem that those people whose attitudes converge, on average, gesture more than those whose attitudes diverge; they talk for longer about green issues and produce a greater number of gestures per minute compared to the divergent group (although there does seem to be considerable individual variation in this overall pattern). In other words, when people have unconscious implicit attitudes that do not match their consciously expressed attitudes, generally speaking they talk less and gesture less (in other words, there is a degree of inhibition in their behaviour). Of course, following Freud, evidence of this attitudinal dissonance within individuals may well be more visible in the detail of *actual examples* of gesture and speech and it is to these we turn.

Bryony is an example of a young consumer who displays a high degree of *convergence* between implicit and explicit attitudes. In this example, she is discussing her own and her family's attitude to recycling. One of the accompanying gestures (gesture 2) in this example is interesting for two reasons: first of all, the gesture adds to the information conveyed in speech by providing additional information about the position of these recycling bins relative to the house (see also Beattie and Shovelton, 1999a, 1999b, who demonstrated that relative position of objects is particularly well encoded by iconic gesture); the gesture seems to indicate that the bins are located to the left of the house (which they were, as it turned out, as we subsequently visited the house). The position of the bins relative to the house is not mentioned in the speech itself, so this is an example of a complementary gesture. This iconic gesture (gesture 2) thus shows her visual thinking about the physical layout of her environment and how that impacts on the process of recycling at her home. The gesture, unconsciously generated, acts as a 'window on the mind' (McNeill 1992), accurately displaying information that is not verbalised in the speech. Gesture 2, as well as displaying relative position, displays the distance of the recycling bins from the house. While Bryony says that the recycling bins are 'just outside', the gesture is a little discrepant to this. It is a gesture that has a relatively long trajectory, suggesting that the bins may be physically further away than she is indicating verbally (again it turns out that in reality they are some distance from the front door, and not 'just outside'). Bryony is someone who says that she is very 'green' in her attitudes; the IAT reveals that her implicit attitudes are also very 'green'. She does recycle, and places things in the recycling bins, and downplays their distance from the front door in her speech, but the distance is not a serious obstacle to this process. Her gesture tells us this.

Bryony

Erm yeah : er : yeah we-we recycle most things in our house like glass :: plastic : paper and [we've got like recycling bins]$_{(1)}$ [just outside]$_{(2)}$ so ::: it's quite easy

[we've got like recycling bins]$_{(1)}$

Gesture 1: *Both arms move from the centre of the body outwards to about a foot apart*

[just outside]$_{(2)}$

Gesture 2: *Left arm moves forward and points to an area to the right of the body*

Clara

Here is another young consumer with convergent explicit and implicit attitudes. In this first gestural sequence she is wrestling with Walker and King's (2008) dilemma about personal responsibility and the importance of individual action to do something about climate change. She says that she should get more actively involved (and of course her explicit and implicit attitudes converging would prime her to actually do something in this regard), rather than leaving it all to others, 'to fight my corner'. She locates these 'others' in her gestural space using a deictic or pointing gesture (gesture 4). Note the late timing of gesture 5. The deictic gesture accompanying 'corner' seems slow in its execution; perhaps it should be coordinated with 'my'. Gestures 6 and 7 indicate that she is aware that she needs to do more in terms of actual behaviour, to 'be the one that gets involved'. Her deictic gesture (gesture 7) completes the utterance. It points back to the same position in the gestural space as gesture 4, and means 'like the others who are currently fighting my corner'.

No that's [true because you think :: that-tha-sh-that's so lazy of me]$_{(3)}$ to think that [some other people are gonna fight my]$_{(4)}$ [corner]$_{(5)}$ ::: [you-you should also be the one that]$_{(6)}$ [gets : you know : that gets involved]$_{(7)}$

[true because you think :: that-tha-sh-that's so lazy of me]₍₃₎

Gesture 3: *Hands are a foot apart, palms facing down. Hands repeatedly push down in a vertical direction*

[some other people are gonna fight my]₍₄₎

Gesture 4: *Index finger of left hand points away from self – signifying 'other people'*

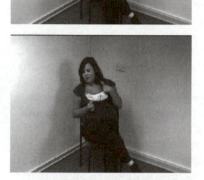

[corner]₍₅₎

Gesture 5: *Index finger of left hand points towards the body – signifying self*

[you-you should also be the one that]₍₆₎

Gesture 6: *Index finger of left hand points away from self again – signifying 'other people'*

[gets : you know : that gets involved]$_{(7)}$

Gesture 7: *Index finger of left hand repeats the gesture of pointing towards the 'other people'*

In this second sequence from Clara, she is trying to find a suitable metaphor to show that change is possible starting with small beginnings. The metaphor she seizes on is that of battery farming and the way that word spread about the unacceptability of this type of farming. Of particular note here is the extent and size of the accompanying gestures. When talking about 'millions of chains of people' the gesture represents these layers and layers of people who are involved. Similarly, gesture 13, used to represent how knowledge and action against battery hens became a 'widespread thing', spreads right out across the gestural space from the centre of the body, extending the left arm as far as it will go, indicating just how widespread and encompassing this issue became. The nature and the scope of her gestures reinforce the content of what she is saying, but add a new dimension through their all-encompassing nature. She believes passionately that such change is possible, and such change is in complete accordance with her explicit as well as her implicit attitude.

Yeah [but then like you said like th:e :: battery he-battery]$_{(8)}$ [hen stuff and the eggs and stuff]$_{(9)}$ [well who started that]$_{(10)}$ :::: [that was just word of mouth and like]$_{(11)}$: [millions of ::: chains of people]$_{(12)}$ and then it just became like : [a w:i:despread]$_{(13)}$ thing didn't it

[but then like you said like th:e ::
battery he-battery]₍₈₎

Gesture 8: *Hands spread out,
moving away from the body with
palms facing upwards, fingers are
spread*

[hen stuff and the eggs and
stuff]₍₉₎

Gesture 9: *Index finger of right
hand points downwards*

[well who started that]₍₁₀₎

Gesture 10: *Hands spread out,
moving away from the body with
palms facing upwards, fingers are
spread*

[that was just
word of mouth
and like]₍₁₁₎

Gesture 11: *Index
finger of left
hand moves
across the body
towards the left
making circular
movement*

[millions of ::: chains of people]₍₁₂₎

Gesture 12: *Hands are raised, making circular movements at descending levels*

[a w:i:despread]₍₁₃₎

Gesture 13: *Left arm moves to the left, away from the body in a sweeping motion*

Gemma

In the gestural sequence below, Gemma is talking about how carbon labelling can guide consumers to the right choice. When talking about the size of carbon footprint labels, she says 'it's just a little thing so that people can see it', but notice the span of the gesture (gesture 15) accompanying 'little thing'. The gesture does not seem to mirror the physical dimensions of the carbon labels available on commercial products. To slightly modify McNeill's description of a growth point example to make the point, 'The gesture and sentence . . . reflected the speaker's conceptualizing of the carbon label as much as the objective carbon label content'

(McNeill 2005: 121). Thus, this speaker through her gestural exaggeration in gesture 15 appears to indicate that the carbon labels should be much more perceptually salient and significant to other people as they are to her with both her explicit and implicit attitudes positive and convergent.

[If it's like what : y'know like in the corner]₍₁₄₎ [like just a little thi:n:g]₍₁₅₎ :: [probably so that people]₍₁₆₎ can ::: [see it]₍₁₇₎ : [if they're gonna]₍₁₈₎ :: but :: not : so it's : that's : all you notice

[If it's like what : y'know like in the corner]₍₁₄₎

Gesture 14: *Makes a circular shape using both hands*

[like just a little thi:n:g]₍₁₅₎

Gesture 15: *Index finger on right hand outlines a circular shape in the air at head height*

[probably so that people]₍₁₆₎

Gesture 16: *Right hand is extended out in front of the body, fingers are spread out, palm facing upwards*

[see it]₍₁₇₎

Gesture 17: *Right hand is raised towards the body*

[if they're gonna]₍₁₈₎

Gesture 18: *Right hand extends out in front of the body again, fingers are spread out, palm facing upwards*

In this second sequence Gemma is talking about the choice dilemma in supermarkets and how if one product had a high carbon footprint and one had a low carbon footprint then she would buy the low-carbon-footprint product even if the low-carbon-footprint product was more expensive (the reason she gives is that she would feel really guilty doing anything else). We know with Gemma that not only is her explicit attitude to low-carbon-footprint products positive but her implicit attitude is as well. Notice how she uses deictic gestures to refer to high- and low-carbon-footprint products in a consistent way by assigning them to a location in gestural space in gestures 19 and 20 (in this case either to her right or left) and then referring back to these concepts within the same location in gestural space in a consistent manner. So when she says that she would 'buy the low', her gesture (gesture 26) moves (and points) to the same location in the gestural space identified in gesture 20. In other words, in both her speech and her gesture she is being perfectly consistent. The unconscious gesture supports rather than contradicts what is being said in the speech of this participant, whose explicit and implicit attitudes match.

Yeah [if it was like really high]$_{(19)}$ [and something was really low]$_{(20)}$ :: [but it was the same product]$_{(21)}$ [er but there was a difference in price]$_{(22)}$ ::: [then I]$_{(23)}$ :::::: [probably still feel really guilty]$_{(24)}$ [about buying the high carbon one]$_{(25)}$ [so I would buy the low]$_{(26)}$

[if it was like really high]$_{(19)}$

Gesture 19: *Right hand moves out to the right of the body, fingers are spread, palms are facing upwards*

[and something was really low]$_{(20)}$

Gesture 20: *Right hand remains extended and the left hand moves out to the left-hand side of the body, fingers are extended, palms are facing upwards*

[but it was the same product]$_{(21)}$

Gesture 21: *Both hands move back into the centre of the body, index fingers on both hands are extended, pointing inwards to an area in the centre of the gestural space*

[er but there was a difference in price]$_{(22)}$

Gesture 22: *Index fingers on both hands then point out, away from the body*

[then I]$_{(23)}$

Gesture 23: *Index finger on the left hand points back towards the body*

[probably still feel really guilty]$_{(24)}$

Gesture 24: *Left hand then extends out, palms are facing upwards, fingers are spread*

[about buying the high carbon one]$_{(25)}$

Gesture 25: *Right hand gestures to the right of the body, palms are facing upwards, fingers are spread. Left hand also moves towards the right with palms facing down*

[so I would buy the low]_{(26)}

Gesture 26: *Both hands flip so that they have moved towards the left-hand side of the body*

Now we can attempt to read the minds of participants whose explicit and implicit attitudes diverge, and perhaps are even dissociated. In the two individuals below the explicit attitudes towards low-carbon-footprint products are positive but their implicit attitudes are much less positive.

Andrew

Andrew has an explicit attitude towards carbon footprint that is positive but an implicit attitude that is less positive, so we can examine how this attitudinal discrepancy between the unconscious and conscious is reflected in his behaviour. One critical point of the interview is when he discusses the moment of choice between high- and low-carbon-footprint products: the example he uses is that of tomatoes. His speech, interestingly, is full of speech disturbances (things like self-interrupted utterances or false starts, e.g. '*and these car-* or ten', parenthetic remarks, e.g. '*you know*', repetition, e.g. '*the,* the lower carbon footprint'), but it is his gestural movements that are particularly revealing here. He locates the high-carbon-footprint tomatoes using his left hand in the left-hand side of his gestural space, the hand has a particular configuration, the back of his hand is outwards away from the centre of the body, the movement is a chopping movement (see gesture 27). He then starts to refer to the low-carbon-footprint tomatoes and his left hand starts to move to the right and point to the right with a different hand configuration (see gesture 28), but he self-interrupts this utterance because he remembers that he hasn't thus far put a number to the high-carbon-footprint products so the hand changes both direction and shape again to identify the high-carbon-footprint tomatoes (as 'ten' in the left-hand side of the gestural space). He then moves his hand to the right, back across the body in the way that he had done previously when referring to the low-carbon-footprint tomatoes to locate the low-carbon-footprint tomatoes in the right-hand side of the gestural space, again using a particular hand shape and hand configuration (back of the hand now pointing up at a 45° angle at this critical point). He also puts a figure to these low-carbon-footprint products, 'six' (see gesture 30)

And then comes the critical movement in the interview: this is the actual movement of choice. Walker and King (2008) tell us that the only way that we can save the planet is by changing our patterns of consumption and our everyday mundane choices. But will we actually do this? Andrew asserts that we will. He says that 'people would be more inclined to go for the low carbon footprint'. You also get the impression that he is including himself in this category; he is really saying 'people like me, people who espouse green attitudes, people who have explicit attitudes (deliberate, conscious, reflexive) that are very pro-low carbon'. But his unconscious hand gesture tells us something quite different here. The hand movement accompanying this critical part of the speech has exactly the same configuration and shape and points to exactly the same location in the gestural space as the gestural movement used earlier to refer to the high-carbon-footprint tomatoes (see gestures 27–29). In other words, we have a mismatch between what he is saying in his speech and what he is saying in his gestural movement. The unconscious movement seems to be reflecting his unconscious implicit attitude and appears to indicate (unambiguously) that his choice would not be the low-carbon-footprint tomatoes but the high-carbon-footprint tomatoes.

What is also fascinating about gesture 31 is its timing with respect to the accompanying speech. In the speech he says, 'then I think people would be more inclined to go for the lower carbon footprint', but notice how the gesture accompanies the words 'then I think'. In other words, that part of the brain which generates gesture movement, working out of his unconscious, has already signified his true thoughts at that point in time. We can read his implicit attitude clearly at this point. No matter how green Andrew says he is, his unconscious gestural movement may tell us a good deal more about his actual behaviour in supermarkets (and also how he thinks others will behave).

One major factor in determining whether attitudes lead to behaviour is our perception of how we think others will behave, and I am convinced that gesture 31 tells us not only about Andrew's implicit attitude, but also about his perception of the subjective norm when it comes to buying low-carbon-footprint products. His gesture tells us that he thinks that very few people will bother with the low-carbon-footprint products (and one can only speculate about his possible reason for this: perhaps he thinks that such products will be significantly more expensive). This therefore lets him off the hook.

Some time ago, McNeill speculated that the kinds of metaphorical gestures that accompany talk can act as 'a window on the human mind', and in this short sequence of bodily movements we get a glimpse into the human mind in action: saying one thing, but beneath the surface of the talk we can see something else going on, something that might hold the clue as to why we all are not doing more to save the planet.

So : yes if-if you've got something you can say : y'know these – [these tomatoes and their carbon footprint is : X]$_{(27)}$ [and : these car]$_{(28)}$ – [or :: [ten] for example]$_{(29)}$ [and these car-these tomatoes and their carbon footprint is six]$_{(30)}$: [then I think]$_{(31)}$ people would be more inclined to go for the :: the lower carbon footprint

[these tomatoes and their carbon footprint is : X]$_{(27)}$

Gesture 27: *Left hand is raised to an area to the left of the body, making a chopping motion*

[and : these car]₍₂₈₎

Gesture 28: *Left hand then starts to move across the body but then stops halfway*

[or :: [ten] for example]₍₂₉₎

Gesture 29: *Left hand sweeps back to the left-hand side of the body and makes a chopping motion*

[and these car-these tomatoes and their carbon footprint is six]₍₃₀₎

Gesture 30: *Left hand sweeps across the body to an area on the right and makes a chopping motion*

[then I think]_{(31)} people would be more inclined to go for the :: the lower carbon footprint

Gesture 31: Left hand sweeps across the body back to an area on the left-hand side, palms are facing upwards, fingers are spread

Sarah

The next example shows something very similar. Each hand is used to represent high- and low-carbon-footprint products, and there is no ambiguity in this person's mind about which is good and which is bad. She refers to them directly as 'the good' and 'the bad' and what she says in her speech explicitly is 'if they were next to each other and it was obvious that one was good and one was bad then you would go for the good one.'

But again, look carefully at the gestural movements that accompany what she is saying. In gesture 35, she uses the left hand opening and closing and raised slightly above and in front of the right hand to signify the good one, that is to say the low-carbon-footprint product. In the next gesture (gesture 36) she uses the right hand with a very similar configuration and hand shape to represent the bad one (i.e. the high-carbon-footprint product), but the critical gesture again is when she talks about the actual moment of choice, she says 'then you'd go for the good one' but the accompanying gesture is executed by the right hand in that part of the gestural space used to represent the bad high-carbon-footprint products. Again this is someone whose explicit attitude is very pro-low carbon but whose implicit attitude is at odds with this, and in gesture 37 you see striking evidence of the unconscious implicit attitude breaking through. This participant may say that she would choose the low carbon option but the unconscious gesture tells us quite a different story.

There is something else that is quite striking about the behaviour of this participant. She repeatedly makes circular movements of the hand when she has displayed her choice in the gestural space, as if some discomfort were associated with the unconscious signalling of the choice. Her intonation is also a little unusual in that it sounds incomplete, again, as if she doesn't want to commit herself. However, in terms of her lexical choices, and the use of specific lexical items, her choice is clear. It is just her unconscious behaviour, and specifically the behaviours over which she has least control (her gestural movements and her intonation), that are sending quite a different message.

I'd probably notice it but : at the same point if yo:u ::: [wanted that product then :: you probably buy it anyway]$_{(32)}$ ::: and obviously [if they were next to each other]$_{(33)}$ :: [and it was obvious]$_{(34)}$ that [one was good and]$_{(35)}$ [one was bad]$_{(36)}$ [then you'd go for ::: [the good one]]$_{(37)}$

[wanted that product then :: you probably buy it anyway]$_{(32)}$

Gesture 32: *Left hand is raised, fingers are spread, hand is lowered using small circular movements*

[if they were next to each other]₍₃₃₎

Gesture 33: *Both hands are raised in front of the chest, hands move backwards and forwards*

[and it was obvious]₍₃₄₎

Gesture 34: *Hands are raised in the centre of the gestural space, palms are facing away from the body, fingers are spread*

[one was good and]₍₃₅₎

Gesture 35: *Left hand then opens and closes, raised slightly above and in front of the right hand*

[one was bad]₍₃₆₎

Gesture 36: *Right hand then moves up so that it is slightly in front of the left hand*

[then you'd go for ::: [the good one]]₍₃₇₎

Gesture 37: *Right hand rises, left hand drops, right hand repeatedly makes smaller circular movements*

There has been very little research until recently on the phenomenology or the experience of holding discrepant explicit and implicit attitudes, one notable exception being an interesting paper by Rydell, McConnell and Mackie (2008), who measured the consequences of holding discrepant explicit and implicit attitudes towards a person ('Bob'). They found that one consequence of this was that when the implicit/explicit discrepancy was greater, dissonance or discomfort was aroused within the individual. In other words, people do not like being in this psychological state. Interestingly, it also had an effect on subsequent information processing such that the more discrepant the implicit and explicit attitudes actually were, the more the individuals then focused on information relevant to the object or concept.

In the gestural analyses we have just seen, we see for the very first time an identifiable behavioural manifestation of a discrepancy between implicit and explicit attitudes that may well help to promulgate a sense of unease or discomfort within the individual concerned. The conscious medium of speech and the unconscious medium of gesture seem to be at odds with one another; perhaps the individuals themselves can sense this behavioural clash and this might be the kind of thing that was reflected in gesture 37. This is a new hypothesis that could be worth testing.

Let me try another one. There is a good deal of evidence (from Festinger 1957 and others) that when people are pressurised to say things that are at odds with their underlying beliefs, they may change their underlying beliefs to align themselves with their behaviour because of cognitive dissonance. If you subscribe to this theoretical point of view, then one way of changing underlying attitudes to the environment is to get people to espouse green attitudes, in effect to get them to tell you how they would make green choices in supermarkets. The theory predicts that (through time) their underlying attitudes would change to match what they have been saying.

But suddenly for the first time we see that things may not be quite as simple as that. When people espouse green attitudes and when this is at odds with their unconscious implicit attitudes, they still have a form of overt communicative behaviour at their disposal (namely, gesture) to communicate their real attitude. Festinger and others analysed only talk itself. In 1957, we did not understand that there is something additional to words that can communicate our ideas and thoughts. So one interesting and important question becomes: when individuals espouse attitudes that are not congruent with their underlying beliefs (because they have been explicitly asked to, or have implicitly felt the social pressure to do this), but where they make these gestural movements that are congruent with their underlying, implicit attitude (like Andrew and Sarah), does this actually prevent a shift in underlying belief state? I think we have a new idea that could run and run in psychology!

The material that we have just been considering is entirely novel research and shows for the first time that when people openly and explicitly espouse green attitudes at odds with those attitudes that they hold unconsciously and implicitly, then we can find behaviour manifestations of this clash. These gesture–speech mismatches are not that common (with a frequency perhaps similar to slips of the tongue), but they do seem to indicate the unconscious slipping through into their observable behaviour.

There are a number of significant implications of this particular bit of research. The first is that what people tell us about their attitude to the environment and what consumer choices they will actually make may, on occasion, be a valuable resource for researchers, but on other occasions people may tell us one thing while their unconscious gestural movement may tell quite a different (and more accurate) story. Therefore, it may be wrong, in research on green issues, to focus *exclusively* on what people say. Explicitly people may want to save the planet, explicitly people may want to appear green, explicitly (and almost certainly) people may want to appear considerate and nice, but implicitly they may care a good deal less. And given that it is these implicit attitudes that direct and control much of our spontaneous and non-reflective behaviour in supermarkets and elsewhere, these are the attitudes that we have to pursue and understand and change.

The second implication is that there does appear to be a dissociation between explicit and implicit attitudes. There does appear to be a conscious mind and an unconscious one. The unconscious mind may not be governed in quite the way that Freud had thought (obsessed with sexual gratification and the libido), but it is there and it does impact on our everyday behaviour. It affects the way we shop, how we talk, how we move, how we gesture, and it can even produce visible signs of discomfort as we speak (with odd, uncertain hand movements and strange, incomplete intonation), sometimes leaving us all a little puzzled: even the green fakers themselves, even me.

The implication of all of this for those wishing to do something about climate change should be clear. It is not sufficient to rely on explicit measures of attitude to low-carbon-footprint products and make assumptions about how easy it will be to change consumer behaviour as a result of providing carbon footprint information (as some others have done: see Downing and Ballantyne 2007). Rather, a significant proportion of individuals have implicit attitudes that are discrepant with their explicit attitudes and this may have significant implications for their 'green' choices. And, on occasion, when you analyse the talk of such individuals you can see the unconscious implicit attitudes of these individuals, so rigidly held and so deeply suppressed, suddenly and unexpectedly revealing themselves as the speaker lays out his or her apparently green agenda. Their sudden appearance seems to surprise everyone, even sometimes the speakers themselves.

13

Taking big risks

So there are green fakers out there, sometimes with odd discrepant body language, who say one thing but believe something different. And in these individuals you can sometimes see their iconic gestural representations go one way as their speech goes another. We need to produce a fundamental change in attitude towards the environment in these people if we are going to change their behaviour towards low-carbon-footprint products (on the assumption that their underlying implicit attitudes are a better predictor of many of the relevant forms of behaviour, including supermarket shopping with all its inherent pressures, than their explicit attitudes). Merely providing carbon footprint information to such individuals (including myself, let's not forget) will currently not necessarily do very much.

But we know that (in principle) this kind of attitude–behaviour change can be done: just look at the change in seatbelt use or the way that people now install smoke alarms as a result of specific explicit attitude change programmes (through public-service advertising: see Lannon 2008, for a description of some important and significant campaigns in this domain). Indeed, some great persuasive messages (well beyond the scope of public-service adverts) have emerged in the past few years with exactly that agenda within the context of sustainability and the environment. Some of these are more than just persuasive messages: they are actually great emotional films, things like *An Inconvenient Truth*, but with the specific, motivated aim of changing how we all feel and act with regard to the environment. But have films like this actually got it right? Do they have any real, demonstrable effect? How should you make people more aware of the risks involved in their own behaviour in terms of global warming? How easy is it to change how people feel inside about core issues like the future of the planet? Indeed, how can you change implicit attitudes at all? I'll start by telling you (from my own experience) how not to.

My whole life people have tried to stop me from taking risks. The first attempt that really made a lasting impression on me and succeeded in changing my behaviour, but not necessarily in the way intended, was a campaign about drugs. I was a teenager at the time growing up in the damp, grey streets of the Belfast of the Troubles. Life was disjointed and fractured. As a teenager my social life was restricted to the streets around me, endless hours of hanging about 'the corner', which was in reality the front of a chip shop with a warm air vent blowing out rancid air that stank of chip fat on cold winter nights. It was a danger-ous and unpredictable place: even the chip shop itself was dangerous, both inside and out. The press called my streets 'murder triangle'. But, of course, I realised even then that there was a life somewhere out there better than this, but it was too far away to glimpse or touch. The world of the *NME*, the *News of the World* with stories about the sordid goings-on of rock stars, images of jeans tucked into green boots, Biba, fast cars. 'Fast cars and girls are easily come by' or 'easy to come by', I can't remember which the pop song said, but not here they weren't. The swings in our local park were chained up on a Sunday lest we enjoy ourselves on the Lord's Day.

It was a Friday in my local youth club when he came to talk to us. We were all boys, I remember that – it was a funny sort of youth club – and we were asked to pull our chairs out into neat rows in front of the speaker. My fingers reeked of coke and crisps. This was an attempt to get at young minds before they were fully formed, before they had been fixed in a pattern; designed to change our behaviour, designed to fit into the evening's youth club activities, designed to warn us of the menace of drugs between table tennis and quarter size snooker, before the dangerous walk home through the streets filled sometimes equally with drunks and terrorists. There was an introduction, then a slide show with images of pills and plants, a glossary of terms, some of which I had heard before, many of which I had not: amphetamine, speed, pep pills, black bombers, dexies, black beauties, black-and-white minstrels, LSD, purple haze, yellow sunshine, blue heaven, sugar cubes, marijuana, dope ('They call it shit here in Belfast,' my friend Colin said helpfully, 'I've never seen it, but I do know that. If you want some, all you have to say is "Can I score some shit?" '), grass, cocaine, coke, Californian Cornflakes. Shit was never mentioned: it was all much more exotic than that.

But to this day I can remember the slides, with shiny red and black pills, white powder as pure as the snow we never saw in our damp streets, exotic plants. From the opening slide I was captivated. It was as if the drugs were jumping off the slides, almost three-dimensional in their appearance. I don't think I blinked once in case I missed something. Things were being revealed to me, to us all: we were all drug virgins, and pop culture virgins. I had a series of agonising shocks of recognition and clarity. 'My friend Jack eats sugar cubes' was no longer a song about a fat teenager with a sugar addiction like fat Albert down the street; 'Purple Haze is in my brain' wasn't a song about pollution and traffic jams and the way that street lights can play odd tricks with your vision when the shipyard was closing and the streets were packed. I was hooked: hooked on the glamour and the glitz, hooked on the terms, with their implicit connotations of something better – 'black beauties', 'yellow sunshine', 'Californian Cornflakes' – hooked on finding the way out from a world where the swings never moved on a Sunday. And when the slides showed close-ups of black bombers, I realised that my rusted bathroom cabinet with the shaky mottled glass door, pinned to the wall in our kitchen (because we didn't have a bathroom or an inside toilet) was full of drugs, full of black bombers, used by my mother as slimming aids.

That night my friends and I took drugs for the first time, and gabbled away outside the chip shop for hours, hardly noticing the smell of chip fat. It probably wasn't that much fun, but we all felt different, separate from everyone else, empowered in a curious sort of way. 'We're on the drugs,' we said to anyone who would listen. And it felt great, dangerous and exciting. It was something that set me apart from the crowd, even though I took a maximum of one tablet at a time, no more or no less than my mother herself, and therefore presumably no more risky. But, of course, it was the context of the taking that gave these small pills their emotional power. For me they were laden with positive emotional connotations, for my mother they were laden with different connotations, connotations of 'putting on the weight', 'and can't get the weight off', 'piling on the pounds, no matter what diet I'm on': connotations of powerlessness and desperation and necessary medical intervention.

This, of course, is just an anecdote, a single case study about the disaffected youth of Belfast one rainy Friday night a long time ago, but it reminds me of the challenge that any attempt to change behaviour faces. Get it wrong and you can, on occasion, get it badly wrong. You can actually make things worse than if you hadn't bothered. The speaker that night with his slides and his spiel did not have a clear understanding of my friends and me, or of our social situation. When you communicate you need a clear model of the audience, their mental state, their needs and aspirations. He had no such model. You have to be able to read other minds. He couldn't. You also need the right approach. He went for a cognitive, rational approach and explained patiently to us that drugs were dangerous. But this meant very little to us. Going for a pint of milk was dangerous where we lived. Telling us that drugs affected the biochemistry of the brain cut no ice either: a night outside the chip shop with the drive-by shootings and then the backfiring cars messed up the biochemistry of your brain. We all knew that. We all had friends who had cracked up after hours spent hanging about the corner doing nothing. None of them needed drugs to help them along. And the presenter underestimated the great emotional pull drugs had for us corner boys: the emotional connotations, London, Biba, long blonde hair in the wind, the Stones, the Who, Led Zeppelin, fashion, rebellion, life in your own hands, not the hands of others;

living dangerously because you wanted to, not because others wanted you to; empowerment, sex, especially sex.

But climate change isn't like drugs, I hear you say. Where's the glamour in global warming? What's so attractive about the submergence of the San Francisco Bay by the Pacific Ocean, or the disappearance of the area around Shanghai which is home to forty million people into the sea due to global warming? Well maybe I'm odd, but I grew up on disaster movies. I can see the human challenge in catastrophe. These films showed me what that challenge was. I can see the glamour in the whole thing. I understand the individual against the elements, the primitive process of survival played out against the most pitiless of backdrops. I close my eyes and I can visualise *The Poseidon Adventure*, where a luxury liner capsises and the passengers trapped in the bowels of the ship have to find their way to safety, with Gene Hackman, Ernest Borgnine and Shelley Winters. I can see Shelley Winters now in my mind's eye, and that panic written on her face in that billowing white dress (I hope that this is the right visual image; I may be transporting images of Shelley Winters from another film into this film: the mind can after all be a very constructive device). I run through *The Towering Inferno* in my mind, where the world's tallest building is destroyed by fire, with Paul Newman, Steve McQueen and Faye Dunaway. Who wouldn't want to be tested like this in the presence of Faye Dunaway?

This might all seem a little perverse but there is a point here, namely that one shouldn't always assume that every individual has the same emotional response to any situation as everyone else, or even the same logical or rational response. When it comes to doing something about climate change we need to think carefully about the psychology underpinning the whole process and reflect that what works for some individuals might not work for everyone. It may be that some people do not understand the logic behind the scientific arguments for climate change (and are too embarrassed to confess this). It may be that some people feel little emotionally about climate change (and are too concerned to confess this). It may be that some people get a slight buzz out of the impending disaster (and are definitely far too sensible to confess this) and what they are relishing is that they, and they alone, will be tested like Gene Hackman, Steve McQueen and Paul Newman and they are waiting with anticipation, and with genuine visceral excitement, for things to deteriorate to give them the right sort of filmic backdrop for their heroic recycling and climate-sensitive actions.

Those great advocates of doing something immediately about climate change do recognise that both rational thought and emotion are core to the persuasion process (although they clearly either haven't met or paid much attention to people like me who might generate some emotional response, but the wrong one, to some of the core messages).

Al Gore's *An Inconvenient Truth* (2006) is probably the best-known and most lauded single communicational message about climate change ever made, obtaining both the Nobel Peace Prize for its author and an Academy Award for Best Documentary Feature. It is in many ways an extraordinary accomplishment: not much more than a lecture with some graphics thrown in, but very engaging and very powerful at a number of levels. It does clearly make you pause to think, and it has produced a strong emotional effect in audiences worldwide.

There are many powerful scenes and arguments. The shots from space of the small blue planet, 'our only home', at the beginning and end of the film (I call this section *the history of the human race*) grab our attention, with both a cognitive and an emotional hook, and feel highly motivating. This particular vista on our earth makes us stop to wonder, to see things differently. The point Gore makes is that the entire history of the human race is contained in this little blue dot, the only home that we will ever have. He presents us with other powerful images and metaphors to understand the nature and the extent of the problem we face. He illustrates the rise in population growth using the time span of his own life to show how the population has changed (clip – *baby boomers*). He tries to manipulate our emotional response with an animation of a polar bear trying to climb onto a floating raft of ice (clip – *drowning polar bear*) that is not thick enough to hold its weight. The message is very concrete and very emotional.

The effects of global warming are illustrated in a scene showing how in the future some of the great coastlines of the world would look if the ice on Greenland were to break up and melt, or even half of the ice on Greenland and half of Western Antarctica, which some scientists see as highly plausible (clip – *rising sea levels*). Al Gore also attempts to explain some of the major paradoxes of global warming, such as the fact that it not only creates more flooding (with an increase in the frequency and intensity of hurricanes, cyclones and typhoons across the globe) but also creates more drought. The images here (clip – *paradox*) are of Lake Chad on the edge of the Sahara Desert.

He is keen to show that this whole thing is a global issue, involving every industrialised and every developing nation. We see film of China's industrial progress (clip – *China*). China has huge coal resources to exploit but it is still using old technologies in coal-burning power stations to meet its rapid economic expansion. The message here is that we all must cooperate to do something about this global problem, including the US and China, the two biggest contributors to greenhouse gas emissions.

Of course, a film like this cannot be all doom and gloom, and Al Gore does try to present an argument and a powerful set of emotional images (images that derive their emotional impact from the implicit message that harm can be undone, that time can be reversed, that we can indeed travel backwards in time 'to pre-1970 levels of emissions') that are essentially empowering for the audience. The argument is basically that every little helps, that we can all do something, that because aspects of consumerism were the root cause of the problem they can be a major part of the solution, and that if we just use more efficient electrical appliances, more fuel-efficient cars, etc., we can do something significant about climate change.

I found a number of these sections of the film (seven in all) to be among the most powerful and provocative (they form the basis for a new study in Chapter 14 and will be described in more detail), in terms of either producing the most compelling arguments or producing the most significant effects on my individual emotional response. In addition, these were some of the segments of the film that had the most enduring effects on what I remembered.

What is interesting about these clips is that they all have particular combinations of rational and emotional force in which sometimes there is more push on one dimension and sometimes more on the other. Furthermore, some of the clips are essentially empowering and enabling, explaining that anything any of us does will make a difference to the whole global issue, but some could potentially have the opposite effect. It might just have been me but the clip about the industrial rise of China and its proliferation of power stations left me, I think, temporarily down. Other clips seemed to lift my spirits.

Of course, most people recognise that emotion is crucial to persuasion, which is why so much of political and economic persuasion, and indeed advertising in general, is aimed at the human emotional system. But how do emotions actually connect to rational thought? And what effect do some of the core parts of Gore's seminal film have on mood state and aspects of thinking? Can we try to be more systematic about the effects of certain parts of this film on how we feel and think, so that we can learn a little more about how a particular mental focus affects us all?

The Gore film is all about showing us what will happen to the world if we do not do something, with interweaving emotional and rational scientific arguments. It is about getting us all to recognise the risks involved in continuing with our current lifestyles. But getting people to recognise the riskiness of their own behaviour can often be a difficult process. We know more generally from endless studies that people are bad at estimating risk. People rely on inferences based on what they remember hearing or observing about the risk in question. In other words, they make a judgement, and this judgement is affected by a number of distinct biases. One type of bias is called the 'availability heuristic', which is that people judge an event as likely or frequent if instances of it are easy to imagine or recall.

One experimental demonstration of the relationship between imageability and risk was carried out by Lichtenstein, Slovic, Fischhoff, Layman and Combs in 1978: they gave subjects the annual death toll of motor vehicle accidents and asked them to estimate the frequencies of forty other causes of death. They found that highly imaginable (and plausible) accidents were judged to cause as many deaths as diseases, whereas diseases actually take about fifteen times as many lives.

You could argue that the availability heuristic makes some sense because frequently occurring events are easier to imagine or recall than infrequent events, but the problem is, of course, that availability is also related to factors unrelated to mere frequency of occurrence. For example, according to some researchers the release of the film *Jaws* meant that people suddenly thought that shark attack was a much more common occurrence than it actually is, on the basis of the graphic depictions of the shark (and its large cavernous mouth) in the film. Therefore, it seems that if you want to emphasise the risks associated with any behaviour you need to make any negative images associated with it as memorable as possible. So the problem in changing behaviour is how to manipulate the availability heuristic so that people will no longer underestimate the risks associated with behaviours that give rise to diseases such as cancer, stroke, asthma or diabetes (hard to form clear images of, hidden, and therefore 'unlikely' to happen).

One way of doing this is to psychologically manipulate the memorability of images. We now know that the most memorable and enduring of all human memories are 'flash-bulb memories', which are hardwired memories designed for human survival and shaped by evolution. These are the kinds of enduring and stable memories that we have if we've ever been in a near-fatal car accident or any other major trauma (see Beattie 2004; Lee and Beattie 1998, 2000): emotional memories, where every single aspect of the scene is encoded apparently for all time by the joint action of two of the most primitive parts of the human brain – the reticular formation and the limbic system. If you have once driven too quickly and you have a near-fatal crash you will have a flashbulb memory of the event – a clear, rich, powerful and enduring image – and you will perhaps (for the first time) realise how dangerous fast driving actually is.

But what really affects whether we have a detailed memory of an event? Like many people I have for a long time been fascinated and depressed by the vagaries of human memory: long before I became a psychologist. Like anyone who has lost a parent early in life I have always longed for vivid memories of my loved one as I wished to relive our days together. I was close to my father and I loved his company, but my memories of those days are weak and disjointed. His voice has no tone or pitch in my memory, and he has no distinctive pattern of movement (how did he walk?), no facial tics that I can recall and his smile is the smile of photographs (that slightly forced smile that shy people do) which I have somehow managed to project back onto his everyday behaviour. But the night he died, and the moment I heard, I can recall with an aching vividness.

My memory of that fateful night is just such a 'flashbulb memory'. This phenomenon was first investigated by two psychologists from Harvard called Roger Brown and James Kulik over thirty years ago. They argued that these memories are hardwired in the human brain because they have a high selection value in evolutionary terms. These memories are triggered by events characterised by a high level of surprise (eliciting a response from the reticular formation) and a high level of 'consequentiality' (eliciting a response from the limbic system). When you have this joint action from two of the most primitive parts of the human brain, this indelible memory is laid down. According to Brown and Kulik the innate basis for this type of memory works as follows.

> To survive and leave progeny, the individual human had to keep his expectations of significant events up to date and close to reality. A marked departure from the ordinary in a consequential domain would leave him unprepared to respond adequately and endanger his survival. The 'Now print!' mechanism must have evolved because of the selection value of permanently retaining biologically crucial, but unexpected events.

The extraordinary thing about flashbulb memories, however, is that it is not necessarily the details of the event itself (or the message) that are recorded for all time, but the circumstances in which you hear the news (you remember who told you, the time, the place, the ongoing activity, etc.). Brown and Kulik argue that for evolutionary survival it is the circumstances that are the crucial thing. You must remember exactly where you were and what you were doing when these surprising and consequential events occurred because, as Brown and Kulik say, 'Nothing is always to be feared or always to be welcomed. It depends. In part, on place.' (1977: 98)

The big psychological question is: can we produce flashbulb memories for events that are not life-threatening and that do not affect personal survival? The answer would appear to be 'yes' because many of us have flashbulb memories for major cultural events such as where we were and who we were with the day we learned of Diana Princess of Wales's death, ten years or more after the event. So flashbulb memories can be elicited by events that are not life-threatening for the individual.

Recently, I wanted to explore how vivid different types of memories are. Are these flashbulb memories more vivid than our happiest personal memories which we reminisce about endlessly? Do we have flashbulb memories about highly surprising and consequential events that do not involve us personally, for example 9/11 or the death of Princess Diana (Brown and Kulik's research would suggest that the answer to this is a very clear 'yes')? My new study was commissioned to mark the launch of UKTV's new history channel 'Yesterday'. I sampled a large number of participants, ranging in age from 18 to 84, using a questionnaire with 32 items: eight asking for information about positive historical memories (e.g. Charles and Diana's wedding), eight investigating negative historical events (e.g. 9/11), eight investigating positive personal events (e.g. your own wedding day) and eight investigating negative personal events (e.g. the death of a loved one), all randomly ordered on the questionnaire.

The results revealed that surprising negative events, as predicted, produced some of the most vivid memories, but interestingly 9/11 produced more vivid memories than even the death of a loved one. 80% of the participants recalled who told them about 9/11, 84% recalled what time it was when they heard, 92% recalled where they were, and 71% recalled their ongoing activity. This represents an extraordinary level of recall nearly eight years after the event. Even more extraordinary was that 71% of our participants (who were old enough) recalled where they were forty-six years after the assassination of John F. Kennedy.

Flashbulb memories tell us something very interesting about how the human brain stores vivid, indelible images and we now know a lot more about the neural mechanisms involved in this process. It also tells us something interesting about ourselves. Why do I, who expressed very little interest in Princess Diana when she was alive, have a flashbulb memory of the moment when I heard about her death? Why did my brain respond to this event as 'consequential' in the way that it did?

Could we ever aim to elicit flashbulb memories carefully and deliberately in order to change behaviour? Could we create flashbulb memories for certain key events involving risky behaviour, so that we affect the availability heuristic and lead people to reappraise the risks associated with their behaviour? For example, could we manipulate the nature of the images that we present to an audience (in public-service advertising or in film) beyond mere shock value to something more meaningful, more personal, more emotional, and more consequential for that individual? This would be a considerable challenge but one that could pay dividends with respect to effecting change in risky behaviour. How can we generate flashbulb memories regarding the effects of smoking on coronary heart disease, for example? How can we generate flashbulb memories for climate change? Did Al Gore, using presumably little more than his own psychological intuition, manage to produce flashbulb memories for what the world will be like if we do not change our behaviour immediately? In other words, has he managed to convince us of the risks associated with global warming, and has he somehow managed to produce indelible images of the effects of climate change?

These are some of the questions that I wanted to answer. I wanted to measure people's responses to the crucial clips in the Gore film in a more systematic way. I wanted to determine how each of these clips affected people emotionally in terms of mood state and also how each of the clips affected rational thinking. I wanted to start to think about the relationship between emotion and thinking in this highly charged area.

More generally, if you want people to think about changing their behaviour, what should you do? Should you frighten them, depress them or elevate their spirits? Should you show them cuddly polar bears trying to clamber onto a raft of ice or the power stations of China belching dark, polluting smoke into the atmosphere? I wanted to begin to understand some of this. This was one set of motivations. The second was to see how good a psychologist Al Gore actually was. He was clearly trying to change our behaviour with his film, by appealing to both our thinking and emotional systems. He wanted to make us aware of the huge risks involved if we did not start to change our behaviour immediately. He wanted to stimulate some of the most primitive parts of the human brain in evolutionary terms to lay down indelible traces. But had he pulled it off?

And what about any other films that have attempted to provide us with shocking detail about the consequences of climate change? I made a point of getting *The Day After Tomorrow* from the local video shop to see how much more likely I thought that an impending climate-change disaster was in my lifetime after watching some graphic scenes of climate-change hell. *The Day After Tomorrow* was the 2004 climate-change disaster movie straight from Hollywood, the opening of which is based loosely on what happened to a group of scientists researching depth of ice on a huge floating ice shelf in Antarctica, called Larsen B. In February 2002 the ice shelf shattered violently, but luckily the Hollywood hero Jack Hall, played by Dennis Quaid, escaped in the nick of time to save (in no particular order) his son trapped in a library in New York (swallowed up by a tsunami generated by climate change that has drowned the city, despite the fact that tidal waves like this cannot be caused by rising temperatures; see Walker and King 2008: 72), his relationship with his son (he was always late picking him up as child because of his busy job as a leading scientist), his relationship with his long-suffering research assistants (loyal to the end, literally in the case of one of them), his relationship with his estranged wife (who cures cancer in children for a living), and the whole of what remains of America itself (although Los Angeles is completely lost, with too much expressed emotion in the film to a series of twisters that have somehow managed to form over land in the film, in contrast to forming over water, which is what science teaches us). As Walker and King comment caustically, 'many of the movie's subsequent events managed to be both exciting and scientifically ludicrous' (2008: 72). The film produces many apparently powerful visual images – the tsunami hitting New York, for example; the Statue of Liberty barely visible out of the flood water; the big freeze – but probably none that affect our estimation of the likelihood of any of this happening.

So what was different from *Jaws* and that huge cavernous reddened mouth that seems to come to mind every time I'm swimming out of my depth in the sea? Perhaps it is the fact that we all feel that some of *The Day After Tomorrow* is implausible and therefore we reject the whole thing. Perhaps it is the scale of the trauma and the human tragedy; perhaps we, as human beings, shut down in situations like this both emotionally and cognitively, and deal with them that way. Perhaps it's just not a very good film. If Los Angeles were to be devastated in the way depicted, then I would normally expect to see a little bit more desperation and helplessness written on the faces of the survivors. Perhaps the film doesn't make us connect with the emotional journey of Jack Hall on his ludicrous physical journey, mainly by foot on snow shoes, through miles and miles back to New York to save his son. So in the case of a film like this we have vivid (but presumably temporary) images that appear not to connect to any perception of risk. *The Day After Tomorrow* certainly produced many surprising images, but perhaps our brains did not see them as consequential in the slightest, because the film was not sufficiently emotionally involving and perhaps even because some of the film was clearly ludicrous (although this would imply that rational thought could influence the formation of these kinds of memories, which remains to be seen).

PART IV

Emotion and thought

PART IV

Emotion and thought

14

An inconvenient truth?

But back to Gore's film. There is clearly something of a paradox about public attitudes to global warming and the public's beliefs about the potential risks entailed. While there seems to be an extraordinary level of agreement and consensus among climate scientists about the seriousness of the risks posed by global warming, the public seem somehow less concerned. In the words of Weber (2006):

> Some [climate scientists] hold this belief so passionately that they go to great lengths to alert the public and politicians to the magnitude of the risks, stepping outside of their typical scientific venues to provide congressional testimony or popular press accounts to trigger action (e.g., Hansen 2004). With some notable exceptions, the concern shown by citizens and governmental officials is smaller and less emphatic than that of climate scientists. (Weber 2006: 1)

Weber's analysis of the possible reasons for this apparent lack of concern on the part of the public is that 'The time-delayed, abstract, and often statistical nature of the risks of global warming does not evoke strong visceral reactions', and he continues: 'The absence of a visceral response on part of the public to the risks posed by global warming may be responsible for the arguably less than optimal allocation of personal and collective resources to deal with this issue' (Weber 2006: 1). In other words, until we can produce a strong emotional response in the public to global warming we may not be able to get people to perceive the real risks involved in climate change. His conclusion is that 'These results suggest that we should find ways to evoke visceral reactions towards the risk of global warming, perhaps by simulations of its concrete future consequences for people's home or other regions they visit or value' (Weber 2006: 1).

In a study published in 2006, Leiserowitz found that, while many Americans believe that climate change is 'real', they consider it 'a low priority relative to other national and environmental issues. These results demonstrate that most of the American public considers climate change a moderate risk that is more likely to impact people and places far distant in time and place' (Leiserowitz 2006: 64). His conclusions, and his call to action, are in many ways similar to those of Weber – 'efforts to describe the potential national, regional and local impacts of climate change and communicate these potential impacts to the public are critical' (Leiserowitz 2006: 64). Like Weber, he believes that the targeting of people's emotional responses is critical in this context because, following Zajonc (1980), he argues that 'affective reactions to stimuli are evoked automatically and subsequently guide rational information processing and judgment. Affect and feelings are not mere epiphenomena, but often arise prior to cognition and play a crucial role in subsequent rational thought' (Leiserowitz 2006: 47).

Thus, there does appear to be an argument that the very nature of the phenomenon of global warming (which is somewhat abstract, statistical and 'scientific') inhibits our emotional response to climate change and constrains our thinking. After all, it is argued, the effects of global warming become visible only over a relatively long time frame. In addition, global warming really does require climate scientists to explain to us what we are witnessing, and they have to persuade us that this is different from some abstract statistical norm, or from what we should be witnessing. And all of this requires us to understand and believe the arguments of scientists ('all with an axe to grind', 'just the latest scientific fad'). For these reasons, we may need global warming to be made much more concrete, much more personal and much more emotionally charged in order to make it a top priority for us all.

It was as if several important people in Hollywood had been listening to some of these arguments, because a number of major films were made at about this time that did specifically attempt to make global warming more real, to add emotional valence to its depiction, and to change how we both thought and felt about the phenomenon. They ranged from the award-winning film by Al Gore released in 2006, *An Inconvenient Truth*, to *Ice Age: The Meltdown* (also released in 2006, and aimed at a slightly younger audience). The goal of Gore's film was to teach us all a valuable and urgent lesson, using something like a lecture mode to accomplish this. But did it work?

One underlying assumption behind movies like this is that, in the words of Kellstedt, Zahran and Vedlitz (2008):

> providing information about global warming – in effect, taking the scientific consensus and popularizing it – will lead to increased public concern about the risks of global warming. The lack of public outcry about global warming, then, is not because the public does not care enough about global warming; it is because they don't know enough about it. The more people know about global warming, the thinking seems to go, the more they will feel personally responsible for it, and also be concerned about it.
> (Kellstedt et al. 2008: 114)

But this, of course, is a very big assumption, especially in the case of something like global warming. There is always the dangerous possibility that the more you know about something as potentially catastrophic as climate change, the less you will feel *personally* responsible for it and the more you may rely on defensive attributions that will shift blame and responsibility elsewhere (see Ross 1977 and Lee and Beattie 1998, 2000 for an analysis of defensive attributions in a somewhat different domain). There is even the strong possibility that you will feel more concerned and worried (primarily an emotional response), but that will not be tied in any way to the intended attributions of responsibility (primarily a cognitive response) or any change in behaviour (personal, political or social) that might actually do something about the impending catastrophe.

Prima facie evidence for this possibility comes from the study by Kellstedt et al. (2008), who carried out a large telephone survey of randomly selected adults in the US in the summer of 2004, questioning them about climate change risk perception (specifically measuring the risks of climate change to personal health, finance, environmental welfare, public health, the economy and environmental integrity), their perceived efficacy to have an effect on climate change and the information they had about climate change (measured simply as a response to the question 'how informed do you consider yourself to be?' on an 11-point scale). Extraordinarily, their results revealed a negative correlation between perceived level of knowledge and concern about global warming, such that 'respondents with higher levels of information about global warming show less concern about global warming' (Kellstedt et al. 2008: 120). In addition, 'as the level of self-reported knowledge increases, the perceived ability to affect global warming outcomes decreases' (Kellstedt et al. 2008: 120). These are very pessimistic results in many ways, because at first sight they would seem to be saying that films like *An Inconvenient Truth*, as brilliant and as informative as they might seem, could easily have the opposite effect on audiences to that expected, leaving people feeling less concerned and less empowered after viewing, which is hardly the intention of the film, or Al Gore! (See also Durant and Legge 2005; Evans and Durant 1995 for some comparable evidence from related domains).

But it is important to consider the possible limitations of the design used in the Kellstedt et al. study and the implications of these for any interpretations of the findings. The design of the study was essentially cross-sectional and correlational. Thus, it offers just a snapshot of a particular point in time (in a popular culture where there is a flux in terms of which climate change issues are currently being discussed and how they are reported and represented). In addition, the correlations of self-reported knowledge and concern about global warming do, by their very nature, allow for multiple interpretations, and do not demonstrate a clear causal connection. Thus, there is always the possibility that respondents who are not very concerned about global warming rate themselves as being very informed about the issues, perhaps because of the pressures of social desirability. The internal implicit reasoning might go something like this (if it were to be made explicit for a moment):

> I am not concerned about global warming: other people
> (including Al Gore and other famous politicians) clearly
> think that I should be; they must think that I am
> particularly ill-informed on these issues and therefore
> do not hold me in high regard. They might even think of
> me as 'stupid'. But they are wrong. 'How informed do
> you consider yourself to be?' That is what the survey
> asks. I consider myself very informed, thank you.

This hypothetical interpretation would essentially reverse the dynamic of the correlation, with 'lack of concern' directing 'self-rated knowledge' (rather than vice versa), and in many ways this allows for a more benign and comfortable interpretation of the empirical findings.

The other major correlation reported in the Kellstedt et al. study, that 'as the level of self-reported knowledge increases, the perceived ability to affect global warming outcomes decreases', is in Kellstedt's terms 'a reasonable finding. Global warming is an extreme collective action dilemma, with the actions of one person having a negligible effect in the aggregate. Informed persons appear to realize this objective fact' (Kellstedt et al. 2008: 120). But their own conclusion here is pessimistic (and biased) in the extreme. Collective action is the joint behaviour of individuals, and without individual behaviour change there will be no collective action. It is not satisfactory to say that informing the public about global warming may reduce self-efficacy, but that is okay, because that is the objective reality. Empowering individuals is the best way of instigating collective behaviour change.

But the effects of information presented in the media on major issues, such as climate change, on public response are clearly a very significant issue in societal terms, and for this reason we decided to employ an experimental design to consider this issue in an attempt to allow more direct interpretations of directionality and ultimately causality. We focused on the Gore film *An Inconvenient Truth* partly because of the brilliance of sections of the film, and partly because of its assumed effect on audiences worldwide. The Gore film has many classic scenes that attempt to manipulate both emotions and social attitudes to achieve their desired ends. It is important to attempt to evaluate the effects of sections of the film on both emotions and cognitions because we now know not only that both are important in terms of behaviour change, but we also know a little more about how these two systems work.

What we know more generally from research in neuroscience on emotion and thinking is that one system (the emotional system) often precedes and directs the other, and that, according to some psychologists, much of so-called rational thought is little more than a post-hoc justification for our behaviour. Some psychologists have even suggested that when we specifically target thinking in which people are apparently making up their mind about certain things, we may be targeting not thinking itself, with implications for subsequent behaviour, but no more than a store of rationalisations for behaviour that is already primed and ready to go as a result of our immediate unconscious emotional reaction (see Beattie 2008).

Antonio Damasio (see Damasio 1994) has been at the centre of much of this new research in neuroscience into how emotion and conscious rational thought connect. His research shows that emotion focuses attention, has a major effect on what we remember and is more closely linked to behaviour than are cognitions (see Walsh and Gentile 2007). But we also now know that in normal people, activation of the emotional system precedes activation of any conceptual or reasoning system (at least in certain domains) and, perhaps as importantly, that the two systems are quite separate. Damasio and colleagues famously showed all of this with a very simple gambling experiment. In front of the participant are four decks of cards; in their hands they have $2000 to gamble with. The task is to turn over one card at a time to win the maximum amount of money: with each card you either win some money or lose some money. In the case of two of the decks the rewards are great ($100) but so too are the penalties. If you play either of these two decks for any period of time you end up losing money. If you concentrate on selecting cards from the other two decks you get smaller rewards ($50) but also smaller penalties, and you end up winning money in the course of the game.

What Damasio found with people playing this game was that after encountering a few losses, normal participants generated skin conductance responses (a sign of autonomic arousal) before selecting a card from the 'bad deck' and they also started to avoid the decks associated with bad losses. In other words, they showed a distinct emotional response to the bad decks even before they had a conceptual understanding of the nature of the decks and long before they could explain what was going on. They started to avoid the bad decks on the basis of their emotional response. Damasio also found that patients with damage to a particular area of the brain called the ventromedial prefrontal cortex failed to generate a skin conductance response before selecting cards from the bad deck, and did not avoid the decks with large losses. Patients with damage to this part of the brain could not generate the anticipatory skin conductance response and could not avoid the bad decks even though they conceptually understood the difference in the nature of the decks before them. In the words of the authors, 'The patients failed to act according to their correct conceptual knowledge' (Bechara, Damasio, Tranel and Damasio 1997: 1294). In other words, Damasio and his colleagues demonstrated that 'in normal individuals, non-conscious biases guide behaviour before conscious knowledge does. Without the help of such biases, overt knowledge may be insufficient to ensure advantageous behavior.' In normal people activation of the emotional system precedes activation of the conceptual system, and we now know that the neural connection between these two systems is located in the ventromedial prefrontal cortex.

More recently, Damasio and colleagues demonstrated

the powerful role of emotions in the generation of moral judgements in that patients with bilateral damage to the same brain region, the ventromedial prefrontal cortex, were more likely to opt for 'heroic' and highly emotional personally aversive responses in a series of moral dilemmas presented to them (Koenigs et al. 2007). Haidt (2001) developed a new model of moral judgement (and evaluative judgement generally) in which *moral judgement* (or *evaluative judgement*) appears in consciousness automatically and effortlessly, but 'Moral reasoning is an effortful process, engaged in after a moral judgment is made, in which a person searches for arguments that will support an already-made judgment' (2001: 181). In other words, we make our mind up pretty quickly and the 'arguments' presented to us may play little role in our judgement except in the subsequent justification of our behaviour to ourselves or others.

This research explains why some behaviour change campaigns work so well. They have targeted the non-conscious biases head-on. Storey (2008) writes that 'Numerous studies have identified that emotional stimuli make far more effective prompts than purely rational arguments when it comes to changing opinions and provoking a response' (2008: 23). The way that the brain is hardwired suggests that this might well be the most appropriate strategy. These non-conscious biases affect behaviour long before we understand the significance of the thing that we are acting towards.

Al Gore tries to manipulate both emotions and cognitions in his film *An Inconvenient Truth* (exactly as Weber and Leiserowitz had recommended). He attempts to make the whole issue of global warming real and concrete for individuals; he attempts to work against the dismissive idea that global warming is something abstract and statistical and just the latest scientific fad, and something that really does not concern *us*. But does it actually work? We considered the effects of a series of extracts from the film on both emotional and cognitive responses. The emotional responses were measured through a mood questionnaire and the cognitive responses were measured through a series of explicit scales relating to both social attitudes and social cognitions. We wanted to know whether extracts from the film impact on our psychological mood in a measurable and reliable way, and how any changes in mood relate to how we think about the film and what we believe that we can do about global warming and the future of our planet.

Seven clips that were identified as being particularly powerful and emotional were picked out from *An Inconvenient Truth*. These were as follows.

Clip 1: China

This clip shows that global warming is a global issue, involving every industrialised and developing nation. We see film of China's industrial progress, its manufacturing industries feeding the world's markets. China's economic advantage is that it has huge coal resources to exploit, but it is still using old technologies in coal-burning power stations to accommodate its rapid economic expansion. The message here is that all countries must unite to do something about this global problem, including or maybe even especially the US and China, the two biggest contributors to greenhouse gas emissions. (According to some reports, China now has the dubious distinction of having overtaken the US as the top emitter of greenhouse gases, but the US still heads the league table in emissions per person, 24.0 tonnes/person compared with 5.0 tonnes/person for China. As many have pointed out, the US and certain Western nations bear 'very significant historical responsibility' for the greenhouse gases already out there; see Walker and King 2008: 209.)

Clip 2: Natural Resources

Of course, a film like this cannot be all doom and gloom, and Al Gore does try to present an argument and a powerful set of emotional images (images that derive their emotional impact from the implicit message that harm can be undone, that time can be reversed, that we can indeed travel backwards in time 'to pre-1970 levels of emissions') that are essentially empowering for the audience in this clip. The argument is basically that every little helps, that we can all do something, that because aspects of consumerism were the root cause of the problem they therefore can be a major part of the solution, and that if we just use more efficient electrical appliances, more fuel-efficient cars, etc. then we can do something significant about climate change.

Clip 3: Small Planet/The History of the Human Race

This is a particularly powerful clip that shows images taken from space of the small blue planet, 'our only home', grabbing our attention with both a cognitive and an emotional hook, and feels very motivating. This particular vista of our earth makes us stop to wonder, to see things differently. The point Gore makes is that the entire history of the human race is contained in this little blue dot: it is our only home.

Clip 4: Paradox

In this clip, Al Gore attempts to explain some of the paradoxes of global warming, such as the fact that it not only creates more flooding (as evidenced by an increase in the frequency and intensity of hurricanes, cyclones and typhoons across the globe) but also creates more drought. The images here are of Lake Chad on the edge of the Sahara Desert. Chad was once one of the biggest lakes in the world but it has now dried to almost nothing, causing major political and social upheaval for the region. Gore painstakingly explains that not only does global warming increase evaporation from the sea, but the higher temperatures also increase soil evaporation, taking all the moisture from the ground.

Clip 5: Drowning Polar Bear

Here, Al Gore explains the effects of global warming on the arctic ice caps with simple graphics, explaining that the ice caps act as mirrors reflecting 90% of the sun's rays. However, as the temperature of the sea rises, the ice caps begin to melt. When the sun's rays hit the ocean instead of the ice caps, 90% of the rays are absorbed, increasing the rate of melting. He tries to manipulate our emotional response with

an animation of a polar bear trying to clamber onto a floating raft of ice; the ice raft isn't thick enough to hold the
weight of the bear. The message, both very concrete and
highly emotional, is that polar bears are now drowning and
that some bears are having to swim up to sixty miles to find
ice.

Clip 6: Population Growth/Baby Boomers

This clip presents us with powerful images and metaphors to
understand the nature and the extent of the problem we face.
Thus, Al Gore illustrates the rise in population growth using
the benchmark of his own life to show how the population
has changed in one baby boomer's lifetime, his own. After the
Second World War the population passed the 2-billion
mark for the first time, but by 2005 it had reached 6.5 billion.
With a single graph he shows how the population would grow
to 9 billion in his own lifetime. He makes the simple but
powerful point that this is extremely worrying given that
it took ten thousand generations for the population to reach
2 billion in the first place.

Clip 7: Rising Sea Levels

The effects of global warming are also illustrated in this clip,
which shows how some coastlines would look in the future
if the ice on Greenland were to break up and melt, or even
half of the ice on Greenland and half of Western Antarctica,
which some scientists see as plausible scenarios for the
future. This scene is introduced with the comment by Al
Gore that Tony Blair's chief scientific adviser has said that
because of the rapid melting seen in Greenland the maps of
the world will have to be redrawn. We then see what the
consequences of global warming and rising sea levels would
be on Florida, the San Francisco Bay, the Netherlands, the
area around Beijing, the land around Shanghai, the area
around Calcutta and Bangladesh and then (one suspects for
many in the US) the clincher – Manhattan, including the site
of the World Trade Center. The graphic that is used is maps
of the regions concerned, with the green becoming blue as

the land is submerged under the rising sea levels. The quantitative message here is stark in the extreme – think of environmental events in the past and their impact on tens of thousands of people; now think about a different order of magnitude altogether. We are told here that these events would involve the death or displacement of one hundred million or more human beings.

Mood questionnaire

In order to measure changes in mood states after watching each clip, a mood questionnaire adapted from the UWIST Mood Adjective Checklist (UMACL) constructed by Matthews, Jones and Chamberlain (1990) was used. The questionnaire was reduced from the original 48 items to 21 items, which were grouped into seven mood categories: happiness, sadness, anger, tension, calmness, energy and tiredness in order to make comparisons about changes in mood state. Responses were noted on a 5-point scale ranging from 1 to 5 where, for example, 1 = not at all cheerful and 5 = extremely cheerful.

The questionnaire was also designed to assess any changes in explicit attitude/social cognitions towards climate change after watching each clip. Thirty statements were designed that could be grouped under five broad categories of explicit attitude/social cognitions, namely: message acceptance, motivation, empowerment, shifting responsibility and fatalism (see Table 14.1). Participants

Table 14.1 Statements used to assess attitudes towards climate change

Category	Statements
Message acceptance	I believe most of what was said in the messages
	I trust most of what was said in the messages
	I believe that the climate is changing
	Climate change is being over-exaggerated (reverse scoring)
	Climate change is a serious issue facing the UK
	I am concerned about climate change

Motivation	I am more concerned about climate change after seeing these messages
	Climate change is a threat to me personally
	I will personally be affected by climate change
	I am prepared to make lifestyle changes to reduce climate change
	I am prepared to change my everyday behaviour to reduce climate change
	I am prepared to do more to help reduce climate change
Empowerment	The UK can make a difference in the fight against climate change
	Climate change is a problem to be solved by my generation
	I can personally help reduce climate change
	Everyone can do their bit in the fight against climate change
	I feel empowered in the fight against climate change
	I am already doing something to help reduce climate change
Shifting responsibility	Climate change is mainly a threat to other countries
	It is the responsibility of other countries, not the UK, to reduce climate change
	Climate change will only affect future generations
	Climate change is a problem to be solved by future generations
	It is not my responsibility to reduce climate change
	I would do more to try and reduce climate change if other people did more as well
Fatalism	I have no control over climate change
	There is no point in me trying to do anything to reduce climate change
	I feel helpless in the fight against climate change
	Climate change is too difficult to overcome
	I feel powerless in the fight against climate change
	Some people do not care about climate change

indicated on a 5-point scale the extent to which they agreed or disagreed with the statement, where 1 = strongly disagree, 2 = disagree, 3 = neither agree nor disagree, 4 = agree, 5 = strongly agree.

Participants were asked to complete the mood questionnaire and the climate change attitudes questionnaire before watching the clips. After completing the questionnaires, participants (in small groups of three or more) were shown the first of the seven clips taken from *An Inconvenient Truth*; the clips were all shown in a random order to the groups. After watching each clip, participants were asked to fill in the mood questionnaire and the climate change attitudes questionnaire again before moving on to the next clip. This procedure was repeated after each clip. The mean responses for the mood questionnaire are shown in Table 14.2 and are represented graphically in Figure 14.1.

Statistical analyses were conducted on the data to reveal whether there were significant changes to mood state after the participants had watched each of the seven clips. The data were split by the midpoint (3) into two categories – high (4/5) versus low (1/2) – for use in the test. The statistical analyses revealed significant changes in mood state changes for *happiness, sadness, calmness* and *tiredness* (but none for *anger, tension* or *energy*), as outlined below.

Participants' ratings of happiness dropped significantly from their baseline happiness levels measured at the start

Table 14.2 **Mean responses on the mood questionnaire**

Category	Overall mean responses							
	Pre	Clip 1	Clip 2	Clip 3	Clip 4	Clip 5	Clip 6	Clip 7
Happiness	3.01	2.56	2.64	2.62	2.16	2.06	1.94	1.13
Sadness	1.37	1.44	1.31	1.25	1.27	1.69	1.31	1.89
Anger	1.25	1.55	1.30	1.33	1.48	1.54	1.68	1.65
Tension	1.60	1.58	1.31	1.35	1.48	1.60	1.63	1.68
Calmness	2.47	2.32	2.35	2.16	2.15	2.10	2.10	1.82
Energy	2.58	2.26	2.17	2.25	2.00	2.06	2.10	2.09
Tiredness	1.78	1.79	1.67	1.63	1.76	1.73	1.46	1.39

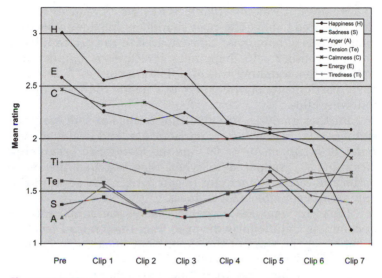

Figure 14.1 **Mean responses on the mood questionnaire.**

of the experiment (mean = 3.01) to their ratings after watching each of the seven clips.

Aside from the pre-viewing condition when levels of happiness were at their highest (mean = 3.01), the *Natural Resources* clip (mean = 2.64) and the *Small Planet* clip (mean = 2.62) left participants with the highest levels of happiness of all the clips, which seemed appropriate as these clips were viewed by the experimenters as 'optimistic clips'. Surprisingly, the China clip came third in the happiness ratings (mean = 2.56). For the remaining clips, however, there seems to be a significant drop in levels of happiness after viewing.

When compared to the *Natural Resources* clip (mean = 2.64), participants were significantly less happy after watching the *Population Growth* clip, mean = 1.94, and after watching the *Rising Sea Levels* clip, mean = 1.13. When compared to the *Small Planet* clip (mean = 2.62), participants were again significantly less happy after watching the *Population Growth* clip, mean = 1.94, and the *Rising Sea Levels* clip, mean = 1.13.

However, it was the *Rising Sea Levels* clip (mean = 1.13) that appeared to have the most significant impact on participants' levels of happiness as, compared to the pre-condition and every other clip, happiness levels were significantly reduced after watching this clip. The *Rising Sea Levels* clip made participants significantly less happy than any of the following clips.

Further analyses revealed that ratings of sadness were significantly higher after participants had watched the *Polar Bear clip*, mean = 1.69, and the *Rising Sea Levels* clip, mean = 1.89, compared to when they watched the more optimistic *Small Planet* clip (mean = 1.25). These were the only significant effects for sadness.

Here the analyses revealed that participants' ratings of calmness significantly dropped from their initial level of calmness at the start of the experiment (mean = 2.47) after watching each of the seven clips.

Analyses here revealed that participants' ratings of tiredness dropped significantly after watching the *Rising Sea Levels* clip (mean = 1.39) when compared to the pre-viewing condition, mean = 1.78, and two of the other clips – the *Paradox* clip, mean = 1.76, and the *Polar Bear* clip, mean = 1.73. There were no other significant differences for this mood state.

The responses for each statement were grouped under the five category headings of message acceptance, motivation, empowerment, shifting responsibility and fatalism, and the mean responses were calculated as shown in Table 14.3. These responses are illustrated in Figure 14.2.

Statistical tests were conducted on the data to test whether there were significant changes to explicit attitudes/ social cognitions after watching each of the seven clips. Changes to attitudes were found for the categories of *motivation, empowerment, shifting responsibility* and *fatalism* but not for the category of *message acceptance*, as detailed below.

Participants' levels of motivation significantly increased after watching each of the clips compared to the pre-viewing condition.

Table 14.3 **Mean responses on the climate change attitudes questionnaire grouped by category**

Category	Overall mean responses							
	Pre	Clip 1	Clip 2	Clip 3	Clip 4	Clip 5	Clip 6	Clip 7
Message acceptance	3.68	3.73	3.79	3.54	3.80	3.82	3.79	3.88
Motivation	3.43	3.55	3.67	3.83	3.72	3.90	3.94	3.98
Empowerment	3.58	3.70	3.84	4.07	3.79	3.80	3.81	3.89
Shifting responsibility	2.69	3.23	2.65	2.18	2.64	2.60	2.69	2.44
Fatalism	2.98	2.95	2.84	2.70	3.02	2.74	2.78	2.88

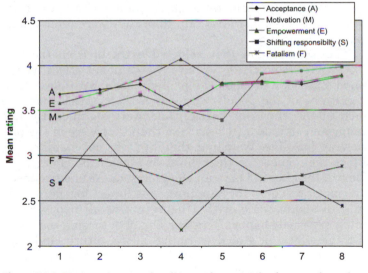

Figure 14.2 **Mean ratings on the climate change attitudes questionnaire coded by category.**

Motivation significantly *decreased* from the pre-viewing levels after watching the *Paradox* clip. After watching the *Polar Bear* clip, levels of motivation were significantly higher than after watching either the China clip or the *Natural Resources* clip. After watching the *Rising Sea Levels* clip, motivation was again significantly higher than after watching either the *China* clip or the *Natural Resources* clip.

After watching the *Population Growth* clip, motivation levels were significantly higher than after watching both the *China* clip and the *Natural Resources* clip. Levels of motivation were also significantly lower after watching the *Paradox* clip than after watching the *China* clip. Compared to the pre-viewing levels, empowerment significantly increased after watching each of the clips.

Levels of shifting responsibility increased significantly after watching the *China* clip compared to the *Rising Sea Levels* clip. None of the other comparisons reached statistical significance. Levels of fatalism were significantly lower after watching the clips compared to the pre-viewing levels of fatalism.

Levels of fatalism were also significantly lower after watching the *Small Planet* clip than they were after watching the *China* clip.

Thus, we found that selected extracts from the film *An Inconvenient Truth* do have a significant effect on the mood state as well as on the explicit social attitudes/social cognitions of the people who watch them. After watching each of the clips our participants were significantly less happy and significantly less calm than they were in the pre-viewing baseline. Watching the clips did not significantly affect their anger or their level of tension or energy, but some of the clips did have an effect on sadness and tiredness. The mean response for happiness was lowest for clip 7 (1.13 on a five-point scale, with 1 being not at all happy and 5 being extremely happy). This is a clip about rising sea levels and the effects of global warming on the future landscape of the world. This clip also produced the highest rating on the sadness scale (a mean of 1.89). In other words, when the impact of global warming is illustrated with its impact on various regions around the world and a figure put to the number of people who would be killed or displaced by this tragedy, this has a more profound impact on the mood of those who watch the film than any of the other messages in any of the other clips. Indeed, happiness level is significantly lower after watching this clip than after watching any of the other six other clips. Interestingly, the *Rising Sea Levels* clip also made people feel significantly less tired (a mean rating of 1.39 compared to a pre-viewing baseline of 1.78). It is as if this clip is jolting people awake, producing a tiredness rating quite different from all of the rest (with the closest response on this mood state being to the *Population Growth* clip, with a mean of 1.46). In other words, extracts from the Al Gore film have a significant impact on our mood but they are not all equal in their force. One clip in particular stands out in producing the most dramatic effects, and this is the clip about rising sea levels.

So now we know something about the effects of these clips on mood state, but what about their effects on our explicit attitudes to climate change and on our social cognitions and attributions about what is to be done about it? The results here were generally extremely positive. There were significant changes in *motivation* to do something to help reduce climate change after watching four of the seven clips. Participants also felt more *empowered* in their fight against climate change after watching each of the seven clips and levels of *fatalism* ('I have no control over climate change'/'Climate change is too difficult to overcome') decreased significantly after watching four of the seven clips.

The only negative effect produced by any of the seven clips was on the dimension of *shifting responsibility* produced by the *China* clip. This clip shows that global warming is indeed a global issue, involving every industrialised and developing nation. It focuses on China's developing progress, its manufacturing industries feeding the world's markets. The clip emphasises that China has huge coal resources to exploit but that it is still using old technology in coal-burning power stations to accommodate its rapid economic expansion. After watching this clip our participants felt a shift of responsibility to other places and to other times ('It is the responsibility of other countries, not the UK, to reduce climate change'/'Climate change is a problem to be solved by future generations'). The mean response on this scale was 3.23 (compared to a pre-viewing baseline of 2.69). The implication of this finding is that it points to the psychological dangers of political scapegoating – when the developed countries highlight the inefficiency of developing countries like China in their manufacturing and in the production of greenhouse gas emissions the message gets through, and even a relatively short clip affects both mood state and what we think can be done about the problem of global warming generally. This should be a lesson to us all.

Compared to the results of Kellsted et al. (2008), whose study seemed to suggest that the more participants knew about global warming, the less concerned they felt about it and the less they felt personally responsible for the problem, this study produced a very different pattern of results. Short informative clips from Al Gore's film *An Inconvenient Truth* clearly provided our participants with a lot of new information about global warming, but instead of disempowering our respondents it had exactly the opposite effect. They felt more motivated to do something about climate change, more able to do something and less likely to think that they had no control over the whole climate change process. The whole process may have been partially directed by their strong emotional response to the clips. These are much more optimistic conclusions, and remind us of the power of strong informative (and emotional) messages on explicit attitude and social cognition generally. There was, however, just one fly in the ointment which reminds us of what to avoid when we seek to communicate about climate change. If we berate China too much for not doing what we currently expect in the West, then we can have a big negative effect on Westerners' own sense of responsibility in the fight against climate change. This could prove to be a terrible own goal if we are not careful.

Of course our new research only attempted to measure and analyse momentary changes in mood state and explicit measures of social attitudes and attributions after watching extracts of a particular film, and an important follow-up study would be to consider how films like that of Al Gore impact on longer-term changes in emotional response when we think about global warming. Would the emotional responses be sustained over longer periods of time? Are those individuals who have watched the film less happy generally about climate change, and how does this impact on their social attitudes to climate change and their social cognitions about global warming in this longer time frame? And what about the significant changes in the 'tiredness' ratings that we observed? Would individuals still feel more 'energised' days and weeks after seeing the film, and how might this translate into any changes in their own behaviour relevant to global warming and climate change? Further, how might this impact on their broader political decision-making processes (and endorsement of 'green' issues in the political domain)? Would they still feel more motivated and more empowered and less fatalistic in this longer time frame, and what would the consequences of this actually be? Can films like this produce such a strong emotional and cognitive response that they make a real difference to how we live our lives? These are important questions with potentially very significant implications for the future of the planet and for us all.

What we do now know is that films like *An Inconvenient Truth* can produce a genuine (and measurable) psychological shock: a shock to both our emotional system and to our cognitive (or attributional) system. But how temporary or enduring this shock really is remains to be seen (and properly investigated in due course).

15

Reaching boiling point?

I was lounging by a pool in Skokie, a northern suburb of Chicago, writing the final chapters of this book on a chlorine-stained notebook. I had just been to the University of Chicago to present my findings at a seminar in the psychology department on the possible dissociation between implicit and explicit attitudes towards the environment and how this might be reflected in unconscious gestures. I had presented at David McNeill's lab, the very epicentre of gestural research, and the talk had been received very well. 'Spectacular,' said David, 'absolutely spectacular.' I was basking in that warm glow of praise that academics so love, and the very warm glow of the Chicago sun. It was 92 degrees, 'a very warm late spring, abnormally warm', the locals were telling me (but without any real concern in their voice), and the weathermen and women, all beautifully turned out with the same small regular features that looked almost artificial, were reminding us that there were, after all, precedents for this sort of weather, fifty or maybe sixty years ago, maybe longer. No reason to be alarmed, they said, and everyone seemed to believe them.

The mood by the pool side was happy and contented; the laughter of children rang out in sharp, shrill bursts. The parents slumped on the sunbeds, the bliss of going nowhere and doing nothing, dozing as the sun heated up towards midday. The parents were occasionally startled by the shrillness of the noise of the children, but eventually they even got used to this noise as the day wore on. It is amazing the way that human beings seem to habituate to almost any situation.

So why aren't we saving the planet? That was the question I posed in the title and I promised to give a psychological perspective on this, to provide some sort of psychological analysis. Well, the single, definitive answer is that there isn't one, but there is a long list of possible reasons.

First there is the feeling of learned helplessness. Everyone agrees that, no matter what we do at the present time, the planet will heat up. It is all now just a matter of degree (although every single degree has huge and severe consequences for the ecology of our planet). In their book, Walker and King (2008) use the analogy of the warming of the oceans. It takes an ocean quite a while to heat up and they say that exactly the same principle holds for global warming, but with a much longer time frame. So, they tell us, the full effects of global warming won't be felt for decades or centuries to come. And here we have the second reason all neatly packaged together with the first. No matter what we do, the earth is going to get hotter and we cannot fine-tune our behaviour to minimise the effects of this because the effects will not be known until after we are long gone. This time lag inhibits behavioural change. One lesson that we did learn from the decades when behaviourism dominated psychology was that for behavioural change to occur the consequences (the rewards or the punishments) must follow the critical behaviours immediately, but in the case of global warming we have a delay of centuries to contend with. So how can we expect behaviour change in the present, in the here and now, today?

Presumably, we will only get this if we have a strong anticipatory negative response each time we engage in certain behaviours. Perhaps the most effective mechanism to promote change here will be the most basic of all – Pavlovian or classical conditioning, through the processes and everyday routines of socialisation. Every time a high-carbon choice is made by a child, the parents will say 'bad' (in a controlled and conscious way, and maybe even in a contrived way) or their facial expression will knit effortlessly and quickly into a frown (and maybe the fast and unconscious frown is the best response falling immediately onto the behaviour in question), and this negative contingency unconsciously processed and stored will drive the behaviour down. In my very first study in psychology (it was my final-year project as an undergraduate), I demonstrated that human verbal conditioning without conscious awareness was indeed possible. Every time that a participant in my experiment paused for 600 milliseconds or more in a story-telling task a light came on which the experimental participant thought was a response from a computer informing them that their story-telling was poor at that point. But the light was contingent only on silent pauses of a certain duration: nothing more, nothing less. And it was odd watching the participants attempt consciously to adapt their story-telling, to make it better in order to keep this small red light off, but at the same time they started repeating words and syllables and using 'ums' and 'ahs' to fill the silent gaps. I had managed to condition them to use fewer 600-millisecond pauses without their conscious awareness, but in order to find time to plan what they were going to say in their speech, they used more and more filled hesitations, almost stammering to keep that light off. Of course, a number of the participants did notice that the light 'sometimes' came on when they weren't saying anything, but incredibly they assumed that the onset of the light reflected the computer's judgement of the clause, or sentence, or the idea before the gap. They did not realise that there was no computer evaluating anything, just a light box activated by periods of silence.

Classical conditioning at times might seem like an odd sort of force, but it is a mechanism that can lay down habits and predispositions to act. It just needs a person or a thing, in the case of my light box (plus a plausible story!), to generate the rewards and punishments. So, as we become more aware of climate change the emotional response of the parents to certain consumer choices could easily be passed on to the next generation. Of course, there is probably something of a generational effect working on this already (although in my research the fact that chronological age did not correlate with implicit attitude, but only with explicit attitude, might give you pause to reflect on this). When I was growing up in Belfast nobody had heard of global warming, and that is why, right at the beginning of this book, Laura gave such an emotional response to all of the lights in my office being on, whereas I had shown no emotional response at all. I had been socialised in earlier decades into a different culture: a culture of materialism if not wealth, a culture where the realisation and expression of personal identity and relationships came often and most easily through the possession and exchange of consumer goods and material objects (and not necessarily grand material objects, as I hope I showed with the story of the fort made by my father). This will have to change, somehow.

The third reason is also bound up with the first two and is connected with the language that Walker and King use in the development of their analogy. They had used the metaphor of the oceans' warming. This may be scientifically appropriate, but in terms of the human psyche it is a dangerous way of thinking. The human mind clings to metaphors to understand the many complexities of the world (see Lakoff and Johnson 1980; Beattie 1988). So it will cling to the oceans' warming as a way of understanding global warming, but a warm luxuriant ocean to bathe in takes all of the sting out of global warming. It almost sounds idyllic; it reminds me of the Seychelles, Belize, Mauritius (an island that I now know). I have an unconscious desire to be in that ocean off Mauritius right now (but ideally not during a cyclone).

So why aren't we saving the planet? We are using the wrong images and metaphors to explain to others what is going on, and we are underestimating the power of the mind to retreat into the metaphor and not see beyond it. Why else aren't we saving the planet? Surely, there are other reasons. We aren't saving the planet because we are essentially optimists who fundamentally believe in the concept of evolution and deep down believe that human beings can adapt to almost any circumstances. Just look at the huge cultural variation across the world in terms of the environments people can live in, I hear people say, from the Inuit of Alaska to the Bushmen of the Kalahari Desert. Surely human beings can learn to adapt to whatever global warming throws at us. It may not prove that life-threatening; after all, Keatinge et al. (2000) found that in Finland temperatures between 14.3 and 17.3°C minimised the number of heat-related deaths, while in Athens the range was between 22.7 and 25.7°C. In other words, in terms of human mortality we cope best with the temperatures we are used to.

Walker and King warned us in their book that global warming is a truly global problem and, they wrote, 'there is a clue in the name', but there are two significant words in that name: 'global', which they commented on, and 'warming', which they didn't. Warming sounds good to me; it sounds pleasant, it sounds gradual, it sounds slow enough to allow human adaptation: you warm soup, you warm up someone you love, and you warm your slippers. Warming sounds cosy and homely, it provokes unconscious images that are far removed from the reality that Walker and King are warning us of, it elicits positive images that provoke our sense of optimism. We need something different: words and metaphors that will shock us, as individuals, and others out of our complacency. And, of course, here lies another major problem: is it up to us or up to those mystical hypothetical others to actually do something?

This is a major issue that reverberates at every level from the most personal to the most political. The industrialised nations blame the emergent nations and vice versa, the Americans blame the Chinese (in particular), the Chinese blame the Americans (in particular). Brazil made the Brazilian proposal (during the 1997 Kyoto negotiations) whereby countries should share the burden of emission cuts according to how responsible they were historically for the problem, such that countries like the UK and Germany with their early industrial revolutions would have to bear a larger share of the cuts than their current levels of emissions would justify. But then you just think that this proposal would have so much more valence and weight if it had come from a country that would itself be penalised, rather than a country like Brazil that would do rather well from it. Human beings can choose to compete or cooperate in the environmental domain, and, of course, only cooperation will work, but every time you have a treaty, a proposal or an argument that seems to be dictated by self-interest, this defines the game as essentially one of competition rather than one of cooperation. It sets up a perceptual set through which we interpret all of the incoming elements and information as essentially to do with competition rather than cooperation, and that is exactly what happens here. Human beings love to compete (although 'love' might not be the right word here), that is our nature and that is why we evolved; now we need to try to do something different. We cannot solve the climate problem by competing against each other, but we could solve the problem by competing against what previous generations have done. That in many senses needs to be the new competition.

So what hope do we have for the future? I think that there are many optimistic threads. In this short book I have not tried to drill into the polar ice caps to send more warnings about our changing world; rather I have tried to drill down into a comparatively small set (in the hundreds rather than the thousands) of human beings living and working and studying in the UK in 2008 and 2009. So what did I find? I found that in terms of explicit attitudes, the vast majority did say that they cared about our planet (as you would expect), but in terms of the unconscious implicit attitudes that people hold, I discovered that they seem to care even more (and this surprised me). Unconsciously, people seem to know that low carbon is good and high carbon is bad; their unconscious automatic responses tell me that.

How widespread is this phenomenon? We don't yet know. When I first presented these findings at a conference at the University of Manchester, one very clever woman working in the retail industry said that she would be a good deal more convinced by my results if I had found a sample of a thousand taxi drivers. She was implying, of course, that taxi drivers wouldn't care in terms of implicit attitude, and maybe not even in terms of explicit attitude (although I just wondered what taxi drivers had ever done to her). But we clearly do need to find out how general these results are. I would suggest that the answer is extremely urgent; we need to know what we are working with in terms of implicit the red condition showed relatively more right frontal activation than those in the green condition or grey condition. Previous research had demonstrated that right frontal activation is associated with avoidance behaviour, and the researchers linked this to another task they used in which participants had to choose either easy or difficult analogy tests. Those who had been exposed to the colour red were more likely to choose the easy task rather than the difficult task.

The conclusion of these researchers was that the perception of the colour red prior to an achievement task has a negative impact on performance, and that 'The findings suggest that care must be taken in how red is used in achievement contexts and illustrate how color can act as a subtle environmental cue that has important influences on behavior' (Elliot, Maier, Moller, Freidman and Meinhardt 2007: 154). This all happens well below the level of conscious awareness; red unconsciously reminds people of the danger of failure and impacts on performance. But that is also why red could work on products with high carbon footprints – it not only elicits attention, but could well lead to consumers avoiding high carbon brands because of the effects on right-sided frontal cortical activity and on avoidance.

So, we can change how products are designed and labelled to provide information about carbon footprint (and provide an unconscious nudge in the right direction), but what else can we do? How can we get them to show the correct emotional response every time they buy a gas-guzzling car, or a high-carbon-footprint light bulb? How can we stop people taking long-haul flights to Mauritius or Chicago? Sometimes giving the basic information that an economy flight to Mauritius for one person uses 1.7 tonnes of CO_2 (calculated on my carbon calculator) might be the wrong information too late. When I discovered this, I felt as if I had just eaten a hamburger in my nearest hamburger joint in Skokie and that it contained 1400 calories (which it probably did). I had the brownie and ice cream to follow because I was now in binge mode, trying to eat my way out of guilt. I had already sullied my body. What difference could some more make?

Al Gore had tried to change our emotional and cognitive

attitude and what the starting point for behaviour change actually is.

I also found that these implicit and explicit attitudes were not correlated across individuals and that there was a degree of dissociation between the two: an individual could be high in one and low in another. This means that there are individuals out there who espouse green attitudes but their implicit unconscious attitudes simply do not correspond to what they say. They say that they are green and that they would always choose low carbon products, and they also say that everyone else would do this as well ('it is the obvious choice'), but from the dark recesses of their brain, deep down in their unconscious mind, they really don't believe it.

This is an important finding because we know that implicit attitude is a much better predictor of the kinds of quick, non-reflective, everyday consumer behaviours than explicit attitude. The proportion of individuals who were significantly higher in explicit than implicit attitude in the present sample was in the region of 13%: this figure could, of course, get much higher as other populations are sampled.

The other implication of this aspect of the research is that you cannot base all your conclusions about people's attitude to the environment on what they actually tell you. They may tell you what they believe (or they may not), but some people, through no fault of their own, have implicit unconscious attitudes that are at odds with what they say. I discovered that when these people are interviewed about their attitudes to the environment they do display behavioural manifestations of this discrepancy between their implicit and explicit attitudes. On occasion (and it is just on occasion), their unconscious attitude is revealed in their unconscious gestures, specifically in the form of gesture–speech mismatches. These mismatches are reminiscent of the slips of the tongue described in detail by Freud over a hundred years previously among the middle- and upper-class inhabitants of Vienna. He also described how the unconscious can break through into actual communicative behaviour. He assumed, at that time, that the communication of thoughts and ideas was only to do with the selection of individual words and their combinations, but we now know, following the pioneering work of David McNeill, that underlying thoughts are realised in speech and unconscious gesture together, at the same time, and the new research described in this book shows that it is the unconscious gestural channel that really gives us a window into the unconscious part of the human mind in action.

So if the results concerning underlying attitudes are relatively optimistic, can we be as optimistic about the likely changes of consumer behaviour? I am not so convinced about this at the present time. Carbon footprint information is now appearing on a number of products and it should be dictating consumer choice, but in my opinion it is fundamentally misconceived. Our detailed analysis (perhaps too detailed from many readers' point of view!) reveals that people do not attend to this information in the normal time frame of supermarket shopping. The carbon footprint information is a mixture of icon, numerical information and text requiring a certain time to process. But it is the unconscious implicit attitudes that drive supermarket shopping, and we must use an iconic representation that appeals to the unconscious mind, which summarises this carbon footprint information. This, in principle, can be done because we do, of course, know that iconic and metaphoric gestures communicate unconsciously but very effectively (of course they have been developed and shaped over hundreds of thousands of years through evolution). But clearly there must be a better way of getting the carbon footprint information to the unconscious mind.

We have used our eye-tracking methodology to test possible alternative formats of the iconic representation, and we found that when colours were used to display a carbon footprint as high or low, our participants attended to this information significantly more (and significantly more quickly) than when the carbon footprint used grams of carbon to represent the size of the footprint. Table 15.1 shows the different formats used on the labels, the total viewing time in seconds and the rank order and position of each label (there were ten participants, each of whom studied for 10 seconds, so the totals in the table represent the time spent fixating the label out of 100 seconds and are effective percentages).

Table 15.1 Number of seconds for which each label was attended to

Label	Total viewing (seconds)	Position
Red circle	48	1
Orange foot	35	2
Red foot	34	3
Orange circle	33	4
Green foot	31	5
360 g	27	6
Green circle	26	7
1500 g	21	8
12 g	20	9

This simple experiment clearly showed that when you are attempting to connect to the minds of individuals, red (and orange) would seem to be good colours for representing carbon footprints, because, at least, they are noticed. And there is another reason why red labels could prove particularly effective here. According to Andrew Elliot and his colleagues from the University of Rochester, if you want to do well in any sort of test you should avoid a red pencil because red impairs task performance, even with brief exposures, as it leads to avoidance. In one experiment he had participants solve anagrams (e.g. anagram: NIDRK; solution: DRINK). What the experimenters varied in this study was the colour of the participant number on the test. What they found was that when the participants were left to solve anagrams for five minutes when the participant number was written in red they solved less than 4½, whereas when the participant number was written in green or black they solved more than 5½. The researchers then looked at the effects of colour on the subsection of an IQ test (analogy subtest) and at the effects of colour of the cover of the test on performance, and again they found that participants in the red condition performed significantly less well than participants in the green condition or white condition; this was also demonstrated with maths performance. Subsequent EEG measures of brain activity revealed that participants in

thinking about the planet through his film *An Inconvenient Truth*, and in my new research I found that part of this film did produce significant emotional effects and it also affected what people thought they could do. This was a clear demonstration that our emotions and cognitions are linked inextricably. People need to understand the risks involved with their current behaviours, and many of the images in the Gore film are oddly indelible. I can close my eyes now and still see the power stations in China and the odd-looking maps of our planet sinking into the sea; how long these images will persist in my mind and how long they will impact on any aspect of my behaviour remains to be determined (they clearly weren't flashbulb memories because details of the circumstances in which I viewed the film have already gone). We clearly do need to do more research to determine what kinds of message provoke the biggest emotional and cognitive changes, and we need this research urgently.

This was my first venture into this area: I have no real idea why deep down inside I commenced this particular journey. It might have been because of Laura Sale; it might not have been. Who knows which way my unconscious mind was directing me? The research was throwing up many more questions than I was answering, that was clear, but questions that I felt deserved an answer.

I had stood the day before in McNeill's lab, talking about my research to a group of young, optimistic and very sharp postgraduate students. I had been a postgraduate at Cambridge, and there I had been socialised into a culture where seminars could be interrupted at any point, during the title if necessary(!), and it was that kind of afternoon that I had at the University of Chicago. It was stimulating and demanding and fun (full of interruptions and heated exchanges), and I realised that there is a lot of talent out there that could see the potential in examining the conscious and unconscious mind at work and how both parts of the human mind might think and feel about the environment.

After the seminar, David McNeill, one of the most charming and creative psychologists I have ever met, walked me around the campus in the sweltering heat, my tee-shirt sticking to me. He showed me the bronze sculpture by Henry Moore that depicts nuclear energy on the site of the world's first nuclear reactor, Chicago Pile-1. Here Enrico Fermi produced the first sustained controlled nuclear reaction in the Manhattan Project. This sculpture was unveiled at 3.36 p.m. on 2 December 1967, twenty-five years to the minute after the actual event. Many visitors think that the sculpture resembles a mushroom cloud, but others just see this as a (somewhat malevolent) human skull. And, of course, it is both: one moment it is a skull, the next it is a cloud. The most destructive force in the world coming from a human mind, the mind encased in its protective skull helmet. I touched the bronze sculpture and it was boiling hot in the Chicago heat in the 90s; I wanted to remember the heat that afternoon (it might just be a stimulus to action), like a foretaste perhaps of what was to come.

But that night I had a strange moment on my nightly run, as I was coming back from Harms Woods just outside Skokie (what an ominous name, I thought later). It was still too hot to run, my brain was working slowly and the road intersections were complex and unfamiliar. As I was crossing the road I looked to my left and I saw it coming round a large bend. I realised that it was not going to be able to stop, even though part of me thought that it was bound to: cars always stop (that is what my sum total of experiences had taught me so far). It was an odd moment that slowed down for ever as the grey bonnet of the car sloughed into me. I somehow managed to stay on my feet, which I suspect was just as well, otherwise the car would have gone over me. I somehow managed to bounce off the car. The driver was very concerned (as were all the other drivers who stopped). He was hesitant and stammering, saying that he wasn't used to runners, out on that road, and he wanted to take me to the hospital for a check-up (I suspect that is the American way), but I insisted on running back to my hotel, my left leg trembling, well beyond my conscious attempts to control it, a strange unfamiliar gait to my run. I had found my flashbulb memory (and I had a black eye for a few weeks as a cue to memory, just in case I needed it).

The problem that human beings have, of course, is that

evolution has prepared us to survive in the here and now, not in a century's time. As an individual human being, one's life can finish at a crossroads in Skokie on a night too hot for this time of year. I am sure that, after last night's run, every intersection I cross will be done with greater care and attention. I have a clear and unmistakeable flashbulb memory of the accident (and of Michael Jackson's death, which I learned about the following day). My learning mechanisms are adapted to allow me to change my behaviour on one trial when it is necessary (see Seligman's classic 1970 paper on this topic), but changing my behaviour for things that may happen in a hundred years' time is a different issue altogether. It is quite simply not how our brains work in terms of implicit and unconscious processes. Richard Dawkins, some time ago, described human beings in terms of the concept of the selfish gene, and perhaps we are all just selfish gene machines trying to survive and trying to get our own genes into the next generation; no more, no less. For us to save the planet we will have to find ways to go beyond some of these natural biological instincts.

But I was pleased that, despite all these negatives, certain messages were getting through and that some psychological change was happening. Our unconscious attitude to the environment seemed to be relatively favourable, even now (I suspect that it would not have been quite so positive in the past, but we have no actual data on this). Our unconscious mind seems to know already that low-carbon-footprint products are good; now we need to make sure that this is translated into actual behaviour.

What will be the stimulus for behavioural change? It will almost certainly not be small essays on the side of products about the size of the carbon footprint or rational arguments with cosy, homely metaphors about global warming (where are my slippers?). We need to recognise that human beings have a conscious and an unconscious mind (Freud at least was right on that, but wrong on the role of the sexual drive and the libido in their subdivision). If we want to change unconscious non-reflective behaviour of the kind that is destroying the planet, we need to communicate more directly with the unconscious mind. It is as simple and as complex as this. But we know that this is at least possible, because we do it every time we unconsciously gesture and other people unconsciously respond to the critical information in the gestures. Of course, recognising that there are two great subsystems in the mind might be one small step in the process of making progress in this area.

So finally, say I was asked to explain in one sentence why we aren't saving the planet (that, after all, is what I was implying in the title – some sort of pithy and short answer), on the basis of the psychological explorations that I have carried out in this book. What would I really say was the answer from a psychological perspective? I think I would probably say something like the following: 'We aren't saving the planet because not enough people care deeply enough, and those who do have failed to understand the minds of those who don't, and how these other minds (and indeed their own) are essentially divided into two great subsystems, each with its own essential instincts and logic. Until we understand more about these instincts and logic and how the two subsystems learn and adapt and direct behaviour, I feel that real cooperative action between human beings will be a major problem in the area of sustainability and climate change.'

Of course, it turned out to be two sentences in the end, and not just one. But I think that this is allowable in the present circumstances. Just think of it as one sentence from each half of my brain, and let's leave it at that. For the moment.

16

Some conclusions and some action plans

Some tentative conclusions

- Global warming clearly requires urgent and cooperative action from us all. As Walker and King (2008) wrote, 'We are all part of the problem, and each of us will need to be part of the solution.' And we, as consumers, can actually do something significant through our everyday behaviours and choices. Carbon labelling can potentially empower consumers to make a significant difference and it could work well with a significant proportion of the population, perhaps with a little more thought about how the footprint is actually represented. Its efficacy depends critically, of course, on underlying attitude.
- Explicit attitudes to low-carbon-footprint products appear to be very positive. Implicit attitudes (a better predictor of actual behaviour in socially sensitive domains such as sustainability, and a better predictor of behaviour when there is any sort of mental or emotional pressure on the decision-making or when decisions are being made under time pressure) also appear to be very positive in the one sample tested in this book. This gives us some grounds for optimism.
- Nevertheless, there is a significant proportion of 'green fakers' out there (the exact proportion to be determined with further research), who explicitly and consciously espouse green attitudes, but whose implicit and unconscious attitude appears to be at odds with their publicly expressed attitude. They may not actually know

what their unconscious attitude actually is (after all, it is unconscious!), and they may live an interesting life in which they are puzzled by many of their everyday behaviours, which might seem perennially at odds with the attitudes that they think they hold.

- This clash between their implicit, unconscious attitude and their explicit attitude can potentially be detected through gesture–speech mismatches, where the form and meaning of their unconscious iconic gestures do not match the accompanying talk, in terms of the core ideas being represented by these two channels of communication. This could be very useful for getting below the surface of what people say in interviews or focus groups.
- Carbon labelling is sometimes just about effective communication (no more, no less), to facilitate the behavioural articulation of the underlying implicit attitude through consumer choice. But we need to think carefully about how we represent and communicate this information.
- However, we will also have to work on, and change, both the explicit and implicit attitudes of significant sections of the population.
- For carbon labelling to work, we must make the carbon label psychologically much more salient than it is at present.
- Visual attention is significantly directed to the carbon footprint only on some products. With certain other products there is very little visual attention to the carbon label in the kind of time frame necessary to make a decision in the typical supermarket shopping experience.
- Many more stages of the communicative process for carbon labelling also need to be considered. We have, so far, considered only the most basic step (visual fixation of the carbon label); we need also to consider the interpretation of the label, how it is emotionally processed, how the information is mentally represented and remembered and the impact of this representation on decision-making.
- We can change how people think and feel about global warming. Sections of Al Gore's film *An Inconvenient*

Truth had a big effect on people both emotionally and in terms of how they thought. They felt *more motivated* to do something about climate change, *more able* to do something and *less likely* to think that they had *no control* over the climate change process after watching some sections of the film. The whole process was almost certainly directed by their strong and significant emotional response to the film. This was, I have to say, a fairly optimistic result.

- So we do now know that with careful thought we can produce a genuine (and measurable) psychological shock to both our emotional system and how we think. But how temporary or enduring this shock is, and how it impacts on implicit attitude (and therefore on many aspects of behaviour), remains to be determined.
- Psychology can provide new insights into the whole process of why (and how) we make changes in our behaviour in response to major issues such as climate change. It may also explain why we often do not.
- Psychology experiments can be painfully slow (and they often throw up as many questions as they answer: see below), but they are necessary as small yet essential building blocks in our creation of knowledge in this important area.

Some action plans

- We need to determine whether implicit or explicit attitudes are better predictors of green consumer behaviour in terms of the purchase of low-carbon-footprint products, and we have immediate plans to use online Implicit Association Tests (IATs) and online explicit attitude measures and relate both of these attitudinal measures to Tesco Clubcard data, as a measure of actual consumer behaviour (Tesco will be funding the research). We should be able to do this with a much larger and much more diverse sample than we have used so far (which will involve sociologists and economists as well, as we all move outside our disciplinary silos).
- It would be really useful to know more about the effects

of implicit/explicit attitudinal dissociation on actual consumer behaviour. When there is a clash between the implicit and explicit attitude within a single individual, what impact does this have on their actual consumer choice and on their processing of information about green issues?

- I want to do more research in order to develop gesture–speech mismatches as a possible reliable indicator of implicit/explicit dissociation. Potentially, we could use novel and innovative behavioural measures like this to get beneath the surface of what people say in interviews and focus groups, in order to infer what their unconscious attitude actually is.

- I also want to consider the effects of implicit/explicit dissociation on aspects of cognitive dissonance and link this to people's responses to persuasive messages about green issues.

- We need to explore the impact of different ways of representing carbon footprint information in the packaging of products on some of the more basic processes of human visual attention (including eye fixation) and to explore the relationship between implicit and explicit attitudes and the direction of visual attention.

- We need to understand more fully the complex relationship between visual attention to carbon footprint information on packaging (using our eye-tracking methodology) and the extraction of the critical information about carbon footprint, and we also need to explore the impact of these processes on actual consumer choice.

- It is important to explore basic visual attentional processes to carbon footprint information when the products are viewed not in isolation (as they were in the research reported in this book) but in the context of other products (as on supermarket shelves).

- We need to explore new ways of changing implicit attitudes using advertising messages of various forms (including, if necessary, using the same kinds of psychological devices that were used to promote such negative behaviours as smoking and alcohol consumption in the decades between the fifties and the eighties!).

- It would be very interesting to learn more about what subliminal 'primes' could be used to alter implicit associations towards low-carbon-footprint products.
- We need urgently to move beyond samples of participants based around a university and to explore implicit and explicit attitudes in very different sections of the population (including taxi drivers if necessary!) to see how general the results we have obtained so far actually are. My guess is that there will be many large and significant groups in the population as a whole in which the implicit attitudes to low-carbon-footprint products are not nearly as positive as the ones that we have so far observed (and that is why more of a change in behaviour has, so far, not been observed).
- It is also crucially important to explore implicit and explicit attitudes in countries such as the USA and China. Global warming is after all a global problem, and we need to know more about how both implicit and explicit attitudes to green issues may vary in an international context, and what the impact of such attitudes might be for policy-makers in these various countries (who will have to take the electorate with them).
- I would also like to analyse politicians from different countries talking about the challenges of global warming to see if there is any evidence of dissociation in their own communicational behaviour. For example, are any gesture–speech mismatches present in their talk when they talk about these sorts of green issues? (See, for example, Beattie 2003 for slightly disturbing evidence of gesture–speech mismatches in politicians' talk, including Tony Blair's speeches, when they are talking about other hot political topics.) If mismatches are present when they talk about global warming, could this potentially have a big impact on how their 'urgent' messages are being received (or not received) by the electorate?
- There is real merit in exploring in much greater detail how emotion directs human behaviour and also how human beings manage to rationalise their actions (perhaps driven primarily by emotional concerns) in a wide

variety of domains, including the broad area of sustainability.

- I would also like to consider the role of metaphor in everyday life and the way that it can direct and shape our thinking, and to think more carefully about the development of new metaphors for communicating the concept of 'global warming' more effectively to the public at large.

- This should be linked to an exploration of the role of metaphor in persuading people to change their consumer habits.

- I think that we also need to look at the effects of cognitive dissonance on promoting change in this area by the usual method of getting people to espouse green issues when their underlying attitude is at odds with this.

- But also we need to see whether the use of contradictory iconic gestures generated during talk can interfere with this whole process of attitude change (driven by dissonance). This could be an important issue.

- We need to understand much more about the maintenance of everyday habits (and ways of disrupting them), and how habits to do with consumption link in to core aspects of the self and self-identity.

- We need to make people more aware of what some of the psychological obstacles might be in preventing behavioural change in the general area of global warming.

- We need to come up with a range of psychological solutions that recognise the essential complexity of human beings (and the conscious and unconscious components of their minds) but manage to have real practical value. We need answers to some of the questions that I have posed here and countless others(!) and we need solutions that will actually work. Or I am sure that it will be too late.

- In the meantime, until we have some answers, we might need to use the following slightly more direct approach shown in the graphic.

And please just note the particular laughing face of the supermarket manager in the background of the graphic (that slightly cruel, scornful laugh that you can illustrate so well

"We're trying to discourage carrier bag use"

(*Private Eye* – August 2008)
Source: www.CartoonStock.com

in cartoons), and the way that he's looking down on the shoppers with their 'Wasteful Bastard' plastic bags. You might not have consciously noticed his expression at first – there is after all so much more going on in the faces of the shoppers in the foreground (the unconscious is a truly wonderful thing) – but my guess is that you probably processed his look unconsciously in the very first few milliseconds of looking at the cartoon, and this may have made the whole thing much funnier (it must have been the supermarket manager's idea to heap scorn on the offenders this way; he's certainly enjoying it).

This green faker certainly laughed until he almost cried.

References

Ajzen, I. (1991) The theory of planned behavior. *Organizational Behavior and Human Decision Processes* 50: 179–211.

Allport, G. W. (1935) Attitudes. In C. Murchison (Ed.), *Handbook of Social Psychology*. Worcester, MA: Clark University Press, pp. 798–884.

Allport, G. (1967) Gordon Allport. In E. Boring and G. Lindzey (Eds.), *A History of Psychology in Autobiography*. New York: Appleton-Century-Crofts, Vol. 6, pp. 3–25.

Allport, G. (1968) An autobiography. In G. Allport (Ed.), *The Person in Psychology: Selected Essays*. Boston: Beacon, pp. 376–409.

Balcetis, E. and Dunning, D. (2006) See what you want to see: Motivational influences on visual perception. *Journal of Personality and Social Psychology* 91: 612–625.

Banaji, M. R. (2001) Implicit attitudes can be measured. In H. L. Roediger, III, J. S. Nairne, I. Neath, and A. Surprenant (Eds.), *The Nature of Remembering: Essays in Honor of Robert G. Crowder*. Washington, DC: American Psychological Association, pp. 117–150.

Banaji, M. R. (2006) Tribute: Anthony G. Greenwald. Harvard University. (Online). www.people.fas.harvard.edu/~banaji/research/mrb_talks/tributes/greenwald.html (Accessed: 9 April 2009).

Banaji, M. R. (2008) Edge: The Implicit Association Test: Mahzarin Banaji and Anthony Greenwald. (Online). www.edge.org/3rd_culture/banaji_greenwald08/banaji_greenwald08_index.html (Accessed: 17 July 2009).

Banaji, M. R. and Hardin, C. (1996) Automatic gender stereotyping. *Psychological Science* 7: 136–141.

Beattie, G. (1988) *All Talk*. London: Weidenfeld and Nicolson.

Beattie, G. (2003) *Visible Thought: The New Psychology of Body Language*. London: Routledge.

Beattie, G. (2004) *Protestant Boy*. London: Granta.

Beattie, G. (2005) *The Body Politic*, News at Ten Thirty, ITV, UK.

Beattie, G. (2008) What we know about how the human brain works. In J. Lannon (Ed.), *How Public Service Advertising Works*. Henley on Thames, UK: World Advertising Research Centre.

Beattie, G., McGuire, L. and Sale, L. (in press) Do we actually look at the carbon footprint of a product in the initial few seconds? An experimental analysis of unconscious eye movements. *International Journal of Environmental, Cultural, Economic and Social Sustainability*.

Beattie, G. and Sale, L. (2009) Explicit and implicit attitudes to low and high carbon footprint products. *International Journal of Environmental, Cultural, Economic and Social Sustainability* 5: 191–206.

—— (under review) How discrepancies between implicit and explicit attitudes on green issues are reflected in gesture–speech mismatches as the unconscious attitude breaks through. *Semiotica*.

Beattie, G., Sale, L. and McGuire, L. (in press) An Inconvenient Truth? Can extracts of a film really affect our psychological mood and our motivation to act against climate change? *Semiotica*.

Beattie, G. and Shovelton, H. (1999a) Do iconic hand gestures really contribute anything to the semantic information conveyed by speech? An experimental investigation. *Semiotica* 123: 1–30.

—— (1999b) Mapping the range of information contained in the iconic hand gestures that accompany spontaneous speech. *Journal of Language and Social Psychology* 18: 438–462.

—— (2000) Iconic hand gestures and the predictability of words in context in spontaneous speech. *British Journal of Psychology* 91: 473–492.

—— (2001) An experimental investigation of the role of different types of iconic gesture in communication: A semantic feature approach. *Gesture* 1(2): 129–149.

—— (2002a) An experimental investigation of some properties of individual iconic gestures that affect their communicative power. *British Journal of Psychology* 93: 179–192.

—— (2002b) What properties of talk are associated with the generation of spontaneous iconic hand gestures? *British Journal of Social Psychology* 41: 403–417.

—— (2005) Why the spontaneous images created by the hands during talk can help make TV advertisements more effective. *British Journal of Psychology* 96: 21–37.

—— (2006) When size really matters: How a single semantic feature is represented in the speech and gesture modalities. *Gesture* 6(1): 63–84.

—— (2009) An exploration of the other side of semantic communication. How the spontaneous movements of the human hand add crucial meaning to narrative. *Semiotica* 176.

Bechara, A., Damasio, H., Tranel, D. and Damasio, A. R. (1997) Deciding advantageously before knowing the advantageous strategy. *Science* 275: 1293–1295.

Berry, T., Crossley, D. and Jewell, J. (2008) Check-out carbon: The role of carbon labelling in delivering a low-carbon shopping basket. (Online). www.forumforthefuture.org.uk/files/Check-out carbon FINAL_300608.pdf (Accessed: 1 January 2010).

Blair, I. V. and Banaji, M. R. (1996) Automatic and controlled processes in stereotype priming. *Journal of Personality and Social Psychology* 70: 1142–1163.

Bosson, J. K., Swann, W. B. and Pennebaker, J. W. (2000) Stalking the perfect measure of implicit self-esteem: The blind men and the elephant revisited? *Journal of Personality and Social Psychology* 79: 631–643.

Bowman, H., Su, L., Wyble, B. and Barnard, P. (2009) Salience sensitive control, temporal attention and stimulus-rich reactive interfaces. In C. Roda (Ed.), *Human Attention in Digital Environments*. Cambridge: Cambridge University Press.

Brown, R. and Kulik, J. (1977) Flashbulb memories. *Cognition* 5: 73–99.

Brunel, F. F., Tietje, B. C. and Greenwald, A. G. (2004) Is the Implicit Association Test a valid and valuable measure of implicit consumer social cognition? *Journal of Consumer Psychology* 14: 385–404.

Bruner, J. S. (1957) On perceptual readiness. *Psychological Review* 64: 123–152.

Bruner, J. S. and Goodman, C. C. (1947) Value and need as organizing factors in perception. *Journal of Abnormal Social Psychology* 42: 33–44.

Chaiken, S. and Trope, Y. (Eds.) (1999) *Dual Process Theories in Social Psychology*. New York: Guilford.

Church, R. B. and Goldin-Meadow, S. (1986) The mismatch between gesture and speech as an index of transitional knowledge. *Cognition* 23: 43–71.

Cienki, A. and Müller, C. (Eds.) (2008) *Metaphor and Gesture*. Amsterdam: John Benjamins.

Clarke, H. M. (1911) Conscious attitudes. *American Journal of Psychology* 32: 214–249.

Cummins, J. (2008) *Great Rivals in History: When Politics Gets Personal*. New York: Metro Books.

Damasio, A. R. (1994) *Descartes' Error: Emotion, Reason, and the Human Brain*. New York: Putnam.

Danesi, M. (1999) *Of Cigarettes, High Heels, and Other Interesting Things: An Introduction to Semiotics*. New York: St Martin's Press.

Dasgupta, N. and Greenwald, A. G. (2001) On the malleability of automatic attitudes: Combating automatic prejudice with images of admired and disliked individuals. *Journal of Personality and Social Psychology* 81: 800–814.

Devine, P. G. (1989) Stereotypes and prejudice: Their automatic and controlled components. *Journal of Personality and Social Psychology* 56: 5–18.

Dollard, J. and Miller, N. E. (1950) *Personality and Psychotherapy*. New York: McGraw-Hill.

Doob, L. (1947) The behavior of attitudes. *Psychological Review* 54: 135–156.

Dostoevsky, F. (1864/1972) *Notes from the Underground*. Harmondsworth, UK: Penguin.

Downing, P. and Ballantyne, J. (2007) Tipping point or turning point? Social marketing and climate change. (Online). Ipsos MORI. www.lowcvp.org.uk/assets/reports/IPSOS_MORI_turning-point-or-tipping-point.pdf (Accessed: 5 January 2009).

Durant, R. F. and Legge, J. S. Jr (2005) Public opinion, risk perceptions, and genetically modified food regulatory policy. *European Union Politics* 6: 181–200.

Elliot, A., Maier, M., Moller, A. C., Freidman, R. and Meinhardt, J. (2007) Color and psychological functioning: The effect of red on performance attainment. *Journal of Experimental Psychology* 136: 154–168.

Ellis, A. and Beattie, G. (1986) *The Psychology of Language and Communication*. London: Lawrence Erlbaum Associates.

Elms, A. C. (1993) Allport's *Personality* and Allport's personality. In K. H. Craik, R. Hogan and R. N. Wolf (Eds.), *Fifty Years of Personality Psychology*. New York: Plenum.

Evans, G. and Durant, J. (1995) The relationship between knowledge and attitudes in the public understanding of science in Britain. *Public Understanding of Science* 4: 57–74.

Faber, M. (1970) Allport's visit with Freud. *The Psychoanalytic Review* 57: 60–64.

Fazio, R., Jackson, J., Dunton, B. and Williams, C. (1995) Variability in automatic activation as an unobtrusive measure of racial attitudes: A bona fide pipeline? *Journal of Personality and Social Psychology* 69: 1013–1027.

Festinger, L. A. (1957) *A Theory of Cognitive Dissonance*. Stanford, CA: Stanford University Press.

Forum for the Future (2007) Retail futures 2022: Scenarios for the future of UK retail and sustainable development. (Online). www.forumforthefuture.org/retail-futures-2022 (Accessed: 5 January 2009).

Forum for the Future (2008) Climate futures: Responses to climate change in 2030. (Online). www.forumforthefuture.org/projects/elimate-futures (Accessed: 5 January 2009).

Freud, S. (1901/1975) *The Psychopathology of Everyday Life*. Volume 5 of the Pelican Freud Library. Harmondsworth, UK: Penguin.

Friese, M., Hofmann, W. and Wänke, M. (2008) When impulses takes over: Moderated predictive validity of explicit and implicit attitude measures in predicting food choice and consumption behaviour. *British Journal of Social Psychology* 47: 397–419.

Friese, M., Wänke, M. and Plessner, H. (2006) Implicit consumer preferences and their influence on product choice. *Psychology and Marketing*, 23: 727–740.

Fromm, E. (1941) *The Fear of Freedom*. London: Routledge.

Gelperowic, R. and Beharrell, B. (1994) Healthy food products for children: Packaging and mothers' purchase decisions. *British Food Journal* 96: 4–8.

Gibson, B. (2008) Can evaluative conditioning change attitudes toward mature brands? New evidence from the implicit association test. *Journal of Consumer Research* 35: 178–188.

Gladwell, M. (2005) *Blink: The Power of Thinking without Thinking*. New York: Little, Brown and Company.

Goldin-Meadow, S. (1997) When gestures and words speak differently. *Current Directions in Psychological Science* 6: 138–143.

Greenwald, A. G. (1990) What cognitive representations underlie attitudes? *Bulletin of the Psychonomic Society* 28: 254–260.

Greenwald, A. G. (1992) New Look 3: Reclaiming unconscious cognition. *American Psychologist* 47: 766–779.

Greenwald, A. G. and Banaji, M. R. (1995) Implicit social cognition: Attitudes, self-esteem, and stereotypes. *Psychological Review* 102: 4–27.

Greenwald, A. G., Klinger, M. R. and Liu, T. J. (1989) Unconscious processing of dichoptically masked words. *Memory and Cognition* 17: 35–47.

Greenwald, A. G., McGhee, D. E. and Schwartz, J. L. K. (1998) Measuring individual differences in implicit cognition: The Implicit Association Test. *Journal of Personality and Social Psychology* 74: 1464–1480.

Greenwald, A. G. and Nosek, B. A. (2001) Health of the Implicit Association Test at age 3. *Zeitschrift für Experimentelle Psychologie* 48: 85–93.

Greenwald, A. G. and Nosek, B. A. (2008) Attitudinal dissociation: What does it mean? In R. E. Petty, R. H. Fazio and P. Brinol (Eds.), *Attitudes: Insights from the New Implicit Measures*. Hillsdale, NJ: Lawrence Erlbaum Associates, pp. 65–82.

Greenwald, A. G., Nosek, B. A. and Banaji, M. R. (2003) Understanding and using the Implicit Association Test: 1. An improved scoring algorithm. *Journal of Personality and Social Psychology* 85: 197–216.

Greenwald, A. G., Poehlman, T. A., Uhlmann, E. and Banaji, M. R. (2009) Understanding and using the Implicit Association Test: III. Meta-analysis of predictive validity, *Journal of Personality and Social Psychology* 97: 17–41.

Greenwald, A. G., Schuh, E. G. and Engell, K. (1990) Ethnic bias in scientific citations. Unpublished manuscript, University of Washington, Department of Psychology.

Gregg, A. P. (2008) Oracle of the unconscious or deceiver of the unwitting? *The Psychologist* 21: 762–766.

Haidt, J. (2001) The emotional dog and its rational tail: A socialist intuitionist approach to moral judgment. *Psychological Review* 108: 814–834.

Hansen, J. (2004) Diffusing the global warming time bomb. *Scientific American* 290: 68–77.

Hausman, A. (2000) A multi-method investigation of consumer motivations in impulse buying behaviour. *Journal of Consumer Marketing* 17: 403–419.

Henderson, J. M. and Ferreira, F. (2004) Scene perception for psycholinguists. In J. M. Henderson. and F. Ferreira (Eds.), *The Interface of Language, Vision, and Action: Eye Movements and the Visual World*. New York: Plenum, pp. 1–58.

Hofmann, W. and Friese, M. (2008) Impulses got the better of me: Alcohol moderates the influence of implicit attitudes toward food cues on eating behavior. *Journal of Abnormal Psychology* 117: 420–427.

Hofmann, W., Gawronski, B., Gschwendner, T., Le, H. and Schmitt, M. (2005) A meta-analysis on the correlation between the Implicit Association Test and explicit self-report measures. *Personality and Social Psychology Bulletin* 31: 1369–1385.

Hofmann, W., Rauch, W. and Gawronski, B. (2007) And deplete us not into temptation: Automatic attitudes, dietary restraint, and self-regulatory resources as determinants of eating behavior. *Journal of Experimental Social Psychology* 43: 497–504.

Holsanova, J., Holmberg, N. and Holmqvist, K. (2008) Reading information graphics: The role of spatial contiguity and dual attentional guidance. *Applied Cognitive Psychology* 22: 1–12.

Ipsos MORI (2008) Carbon labelling – just a load of hot air? (Online). www.ipsosmori.com/researchspecialisms/publicaffairs/reputationcentre/carbon.ashx (Accessed: 9 April 2009).

Jacoby, L. L., Kelley, C. M., Brown, J. and Jasechko, J. (1989) Becoming famous overnight: Limits on the ability to avoid unconscious influences of the past. *Journal of Personality and Social Psychology* 56: 326–338.

Jefferson, G. (1990) List-construction as a task and a resource. In G. Psathas (Ed.), *Interaction Competence*. (pp. 63–92) Washington, DC: University Press of America.

Karpinski, A. and Hilton, J. L. (2001) Attitudes and the Implicit Association Test. *Journal of Personality and Social Psychology* 81: 774–788.

Karpinski, A. and Steinman, R. B. (2006) The single category Implicit Association Test as a measure of implicit social cognition. *Journal of Personality and Social Psychology* 91: 16–32.

Karpinski, A., Steinman, R. B. and Hilton, J. L. (2005) Attitude importance as a moderator of the relationship between implicit and explicit attitude measures. *Personality and Social Psychology Bulletin* 31: 949–962.

Keatinge, W. R., Donaldson, G. C., Cordiolli, M., Martinelli, M., Kunst, A. E., Mackenbach, J. P., et al. (2000) Heat related mortality in warm and cold regions in Europe: Observational study. *British Medical Journal* 312: 670–673.

Kellstedt, P. M., Zahran, S. and Vedlitz, A. (2008) Personal efficacy, the information environment, and attitudes toward global warming and climate change in the United States. *Risk Analysis* 28: 113–126.

Kendon, A. (2004) *Gesture: Visible Action as Utterance*. Cambridge: Cambridge University Press.

Kita, S. (2000) How representational gestures help speaking. In D. McNeill (Ed.), *Language and Gesture*. Cambridge: Cambridge University Press, pp. 162–185.

Koenigs, M., Young, L., Adolphs, R., Tranel, D., Cushman, F., Hauser, M. and Damasio, A. (2007) Damage to the prefrontal cortex increases utilitarian moral judgements. *Nature* 446: 908–911.

Lakoff, G. and Johnson, M. (1980) *Metaphors We Live By: Metaphorical Structure of the Human Conceptual System*. Chicago: University of Chicago Press.

Landy, D. and Sigall, H. (1974) Beauty is talent: Task evaluation as a function of the performer's physical attractiveness. *Journal of Personality and Social Psychology* 29: 299–304.

Lannon, J. (Ed.) (2008) *How Public Service Advertising Works*. Henley on Thames, UK: World Advertising Research Centre.

Leahy, T. (18 January 2007). Tesco, carbon and the consumer. (Online). Tesco. www.tesco.com/climatechange/speech.asp (Accessed: 9 April 2009).

Lee, V. and Beattie, G. (1998) The rhetorical organization of verbal and nonverbal behaviour in emotion talk. *Semiotica* 120: 39–92.

Lee, V. and Beattie, G. (2000) Why talking about negative emotional experiences is good for your health: A microanalytical perspective. *Semiotica* 130: 1–81.

Leiserowitz, A. (2006) Climate change risk perception and policy preferences: The role of affect, imagery, and values. *Climate Change* 77: 45–72.

Lichtenstein, S., Slovic, P., Fischhoff, B., Layman, M. and Combs, B. (1978) Judged frequency of lethal events. *Journal of Experimental Psychology: Human Learning and Memory* 4: 551–578.

Louw, A. and Kimber, M. (2007) The power of packaging. (Online). The Customer Equity Company. www.tnsglobal.com/_assets/files/The_power_of_packaging.pdf (Accessed: 9 April 2009).

McNeill, D. (1985) So you think gestures are nonverbal? *Psychological Review* 92: 350–371.

McNeill, D. (1992) *Hand and Mind: What Gestures Reveal About Thought.* Chicago: University of Chicago Press.

McNeill, D. (2000) *Language and Gesture.* Cambridge: Cambridge University Press.

McNeil, D. (2005) *Gesture and Thought.* Chicago: University of Chicago Press.

McNeill, D., and Duncan, S. (2000) Growth points in thinking for speaking. In D. McNeill (Ed.), *Language and Gesture.* Cambridge: Cambridge University Press.

Maison, D., Greenwald, A. G. and Bruin, R. (2001) The Implicit Association Test as a measure of consumer attitudes. *Polish Psychological Bulletin* 2: 61–79.

Maison, D., Greenwald, A. G. and Bruin, R. H. (2004) Predictive validity of the Implicit Association Test in studies of brands, consumer attitudes and behavior. *Journal of Consumer Psychology* 14: 405–415.

Matin, E. (1974) Saccadic suppression: A review. *Psychological Bulletin* 81: 899–917.

Matthews, G., Jones, D. M. and Chamberlain, A. G. (1990) Refining the measurement of mood: The UWIST Mood Adjective Checklist. *British Journal of Psychology* 81: 17–42.

Meringer, R. and Mayer, K. (1895) *Versprechen und Verlesen.* Stuttgart, Germany: Goschensche.

Müller, G. E. and Pilzecker, A. (1900) Experimentelle Beiträge zur Lehre vom Gedächtnis. *Zeitschrift für Psychologie, Erganzungsband* 1: 1–128.

Myers, D. G. (1987) *Social Psychology* (2nd edn). New York: McGraw-Hill.

Nelson, K. (2007) *Young Minds in Social Worlds: Experience, Meaning, and Memory.* Cambridge, MA: Harvard University Press.

Nietzsche, F. (1871/1962) *Twilight of the Idols.* Harmondsworth, UK: Penguin.

Nosek, B. A. (2005) Moderators of the relationship between implicit and explicit evaluation. *Journal of Experimental Psychology: General* 134: 565–584.

Nosek, B. A. (2007) Understanding the individual implicitly and explicitly. *International Journal of Psychology* 42: 184–188.

Nosek, B. A. and Banaji, M. R. (2002) (At least) two factors moderate the relationship between implicit and explicit attitudes. In R. K. Ohme and M. Jarymowicz (Eds.), *Natura Automatyzmow*. Warsaw: WIP PAN and SWPS, pp. 49–56.

Nosek, B. A., Banaji, M. R. and Greenwald, A. G. (2002) Harvesting implicit group attitudes and beliefs from a demonstration website. *Group Dynamics* 6: 101–115.

Nosek, B. A. and Hansen, J. J. (2008) The associations in our heads belong to us: Searching for attitudes and knowledge in implicit evaluation. *Cognition and Emotion* 22: 553–594.

Nosek, B. A. and Smyth, F. L. (2007) A multitrait–multimethod validation of the Implicit Association Test: Implicit and explicit attitudes are related but distinct constructs. *Experimental Psychology* 54: 14–29.

Olson, M. A. and Fazio, R. H. (2004) Reducing the influence of extra-personal associations on the Implicit Association Test: Personalizing the IAT. *Journal of Personality and Social Psychology* 86: 653–667.

Osgood, C. E. (1957) A behavioristic analysis of perception and language as cognitive phenomena. In *Contemporary Approaches to Cognition*. Cambridge, MA: Harvard University Press, pp. 75–118.

Ramchandani, N. (4 September 2006) A few words that say so much. *The Guardian*.

Rayner, K. (1998) Eye movements in reading and information processing: 20 years of research. *Psychological Bulletin* 124: 372–422.

Ross, L. (1977) The intuitive psychologist and his shortcomings. In L. Berkowitz (Ed.), *Advances in Experimental Social Psychology*. New York: Academic Press, pp. 173–220.

Rundh, B. (2005) The multi-faceted dimension of packaging: Marketing logistic or marketing tool? *British Food Journal* 107: 670–684.

Rydell, R. J., McConnell. A. R. and Mackie, D. M. (2008) Consequences of discrepant explicit and implicit attitudes: Cognitive dissonance and increased information processing. *Journal of Experimental Social Psychology* 44: 1526–1532.

Scarabis, M., Florack, A. and Gosejohann, S. (2006) When consumers follow their feelings: The impact of affective or cognitive focus on the basis of consumers' choice. *Psychology and Marketing* 23: 1005–1036.

Seligman, M. E. P. (1970) On the generality of the laws of learning. *Psychological Review* 77: 406–418.

Silayoi, P. and Speece, M. (2004) Packaging and purchase decisions. *British Food Journal* 106: 607–608.

Stern, N. H. (2006) *The Economics of Climate Change: The Stern Review*. Cambridge: Cambridge University Press.

Storey, R. (2008) *Initiating Positive Behaviour: How Public Service Advertising Works*. Henley on Thames, UK: World Advertising Research Centre.

Swanson, J. E., Rudman, L. A. and Greenwald, A. G. (2001) Using the Implicit Association Test to investigate attitude–behavior consistency for stigmatized behavior. *Cognition and Emotion* 15: 207–230.

Uttal, W. R. and Smith, E. (1968) Recognition of alphabetic characters during voluntary eye movements. *Perception and Psychophysics* 3: 257–264.

Vantomme, D., Geuens, M., De Houwer, J. and De Pelsmacker, P. (2005) Implicit attitudes toward green consumer behavior. *Psychologica Belgica* 45: 217–239.

Vygotsky, L. S. (1986) *Thought and Language*. Edited and translated by E. Hanfmann and G. Vakar, revised and edited by A. Kozulin. Cambridge, MA: MIT Press.

Walker, G. and King, D. (2008) *The Hot Topic: How to Tackle Global Warming and Still Keep the Lights On*. London: Bloomsbury.

Walsh, D. and Gentile, D. (2007) Slipping under the radar: Advertising and the mind. In L. Riley and I. Obot (Eds.), *Driving It In: Alcohol Marketing and Young People*. Geneva, Switzerland: WHO.

Weber, E. U. (2006) Experience-based and description-based perceptions of long-term risk: Why global warming does not scare us (yet). *Climate Change* 77: 103–120.

Wilson, T., Lindsey, S. and Schooler, T. Y. (2000) A model of dual attitudes. *Psychological Review* 107: 101–126.

Wundt, W. (1900) *Völkerpsychologie*, I, Part I. Leipzig, Germany: Engelmann.

Young, S. (2003) Winning at retail: Research insights to improve the packaging of children's products. *Young Consumers* 5: 17–22.

Zajonc, R. B. (1980) Feeling and thinking: preferences need no inferences. *American Psychologist* 35: 151–175.

Zeithaml, V. A. (1988) Consumer perceptions of price, quality, and value: A means–end model and synthesis of evidence. *Journal of Marketing* 52: 2–22.

Index